1st International Conference on Computing, Technology, Science and Management in Sports (ICoTSM 2017)

AA006382

Journal of Physics: Conference Series Volume 1020

Kuching, Malaysia
25 – 27 November 2017

ISBN: 978-1-5108-6402-3
ISSN: 1742-6588

Printed from e-media with permission by:

Curran Associates, Inc.
57 Morehouse Lane
Red Hook, NY 12571

Some format issues inherent in the e-media version may also appear in this print version.

Copyright© (2017) by the Institute of Physics
All rights reserved. The material featured in this book is subject to IOP copyright protection, unless otherwise indicated.

Printed by Curran Associates, Inc. (2018)

For permission requests, please contact the Institute of Physics at the address below.

Institute of Physics
Dirac House, Temple Back
Bristol BS1 6BE UK

Phone: 44 1 17 929 7481
Fax: 44 1 17 920 0979

techtracking@iop.org

Additional copies of this publication are available from:

Curran Associates, Inc.
57 Morehouse Lane
Red Hook, NY 12571 USA
Phone: 845-758-0400
Fax: 845-758-2633
Email: curran@proceedings.com
Web: www.proceedings.com

Table of contents

Volume 1020

1st International Conference on Computing, Technology, Science and Management in Sports (ICoTSM) 2017

25–27 November 2017, Kuching, Sarawak, Malaysia

Accepted papers received: 1 May 2018
Published online: 1 June 2018

Preface

Papers

IOP Conf. Series: Journal of Physics: Conf. Series **1020** (2018) 011001 doi:10.1088/1742-6596/1020/1/011001

Preface/Introduction

The quest for knowledge has become part of national main agenda. In line with the country's aspiration to enhance research and development in the country, UniMAP has been proactively involved and initiating programs and events, which would encourage high impact research encompassing various fields of expertise.

The Centre of Excellence Geopolymer and Green Technology (CEGeoGTech) has been playing a significant role in this quest by taking up this challenge. Thus, one of the initiatives grabbed by the Green Advanced Computing and Technology (GREAT), CEGeoGTech is in organizing this conference itself, which provides a fine platform for knowledge sharing and research ideas exchange in the field of Big Data, Cloud Computing, Green and Sustainable Computing, Technology Science and Management in Sports.

This is indeed an opportune moment to strengthen the bond among researchers and professionals from the academics, the industries and the public sectors in sharing new ideas, and in collaborating future research activities. The importance of interaction and networking among peers are the inspiration of this joint conference. I would like to congratulate the organizing committee and all parties involved for their sheer dedication and tireless effort in making this joint conference a great success. I hope that all participants will have a productive and pleasant conference.

Content from this work may be used under the terms of the Creative Commons Attribution 3.0 licence. Any further distribution of this work must maintain attribution to the author(s) and the title of the work, journal citation and DOI.
Published under licence by IOP Publishing Ltd

IOP Conf. Series: Journal of Physics: Conf. Series **1020** (2018) 011002 doi:10.1088/1742-6596/1020/1/011002

Peer review statement

All papers published in this volume of *Journal of Physics: Conference Series* have been peer reviewed through processes administered by the proceedings Editors. Reviews were conducted by expert referees to the professional and scientific standards expected of a proceedings journal published by IOP Publishing.

Content from this work may be used under the terms of the Creative Commons Attribution 3.0 licence. Any further distribution of this work must maintain attribution to the author(s) and the title of the work, journal citation and DOI.
Published under licence by IOP Publishing Ltd

The Influence of Compression Stocking on Jumping Performance of Athlete

M N Salleh[1], H M Lazim[1], H Lamsali[1], A F Salleh[2]

[1] Technology and Supply Chain Institute, School of Technology Management and Logistics, Universiti Utara Malaysia, 06010 Sintok, Kedah
[2] School of Mechatronic Engineering, University Malaysia Perlis, 02600 Arau, Perlis

najib@uum.edu.my

Abstract. Evidence of compression stocking effectiveness are mixed, with some researchers suggests that the stocking can enhance performance while others dispute the finding. One of the factors that are thought to cause the mixed results is level of pressure used in their studies. This research had organized a test on fourteen athletes. Their body was scanned and a customized compression stocking which can exert pressure correspond to the intended one was developed. An experiment was conducted to measure the effect of wearing compression stocking on jumping performance. The results show mixed outcomes. For the female athlete, there is a significant difference between wearing and not wearing compression stocking ($p<0.05$) on knee power. However, there is no significant difference for male athletes whether wearing or not.

1. Introduction

A compression stocking is a garment that is used to enhance the performance of athlete, to prevent injury and to improve their recovery process. It works by enhancing lactate removal, reducing muscle oscillation and positively influencing psychological factors. Although there are journal papers have discussed their effectiveness on enhancing performance (i.e. [1] [2][3][4]), others researchers dispute the finding (i.e. [5][6]). [7] suggests that the level of pressure used in the experiments as the cause for the variation. This is because all the conducted experiments used compression stocking that is made by mass production method. The drawback of the stocking developed by this method is the level of pressure that the compression stocking applies on the leg is sometimes too high or too low because it was made based on common size. The problem can affect the performance of the stocking and consequently to the athlete. Fitting is important in a compression stocking, as it determines the pressure for effective performance. Even though there is no agreement between researchers as to the exact pressure that needs to be applied by a compression stocking, most believe that pressures between 20 mmHg to 40 mmHg are required [7]. [8] had developed a method to design a customized compression stocking using body scanner. This research used the method to develop a customized graduated compression stocking (Figure 1) and investigated the effect of wearing it on jumping performance.

Content from this work may be used under the terms of the Creative Commons Attribution 3.0 licence. Any further distribution of this work must maintain attribution to the author(s) and the title of the work, journal citation and DOI.
Published under licence by IOP Publishing Ltd

IOP Conf. Series: Journal of Physics: Conf. Series **1020** (2018) 012001 doi:10.1088/1742-6596/1020/1/012001

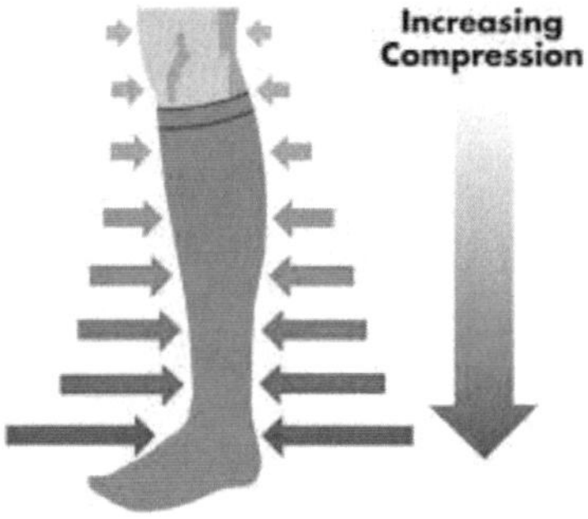

Figure 1: Graduated Compression Stocking (GCS)

2. Methodology

Fourteen active and healthy male and female subjects (7 male and 7 female) were selected from Universiti Utara Malaysia (UUM). They were scanned by using a three dimensional $[TC]^2$ NX12 body scanner. Then, a compression socks for the scanned body parts was constructed. The socks were designed to exert 37.5 mmHg (5000 Pa) at the ankle and gradually reduce to 22 mmHg (3000 Pa) at knee. Method to develop the compression stocking is explained in [8]. Basically, a 3D horizontal type body scanner is used to extract 3D (x, y, z coordinates) raw data of the athlete's body. Then, the data is processed and a 3D compression stocking model is developed. A 2D stocking's pattern was generated from the 3D model. The pattern model was printed out and compression stockings with an exact shape and size were cut from the fabric with margins for sewing purposes. Before the pattern was cut, the paper patterns were aligned on the fabric so that the fabric pattern was mostly made from straight yarn. The pattern has then been sewn together to make the compression stockings.

The exercise experiments were conducted at Biomedical Laboratory, School of Mechatronics Engineering, University Malaysia Perlis (UniMAP). Before the experiments, the subject's anthropometric data were collected to make sure they fulfill the criterion needed. Kinematics data throughout the trial were collected using 5 Oqus cameras (Qualisys Motion Capture System, Gothenburg, Sweden) sampling at 120 Hz. 15 numbers of passive-reflective markers were placed on subject's both lower limbs. The subjects were briefly explained about the purposed of the study and experimental procedure according to the protocol. Before each trial, the subjects were asked to perform an adequate stretching exercise and warming up. Then, a practice session of the jumping exercise was held for each subject prior to experiment until they feel comfortable with the movement. The practice is to ensure that the jumping exercise is conducted in a correct manner. A two seconds static trial was recorded with the subject standing still to create an anatomical reference frame for each limb segment. Markers placed on bony landmarks (medial and lateral malleolus, lateral and medial tibial plateau, lateral and medial femoral condyles, greater trochanter, anterior and posterior superior iliac spine, and the iliac crest) during the static trial were used to establish the tibial, femoral and pelvic anatomic coordinate systems. In dynamic trials, some of the markers used in the static trial were removed except markers at the foot and cluster markers were placed at the thigh and leg for motion tracking purposes. Subjects were asked to jump three times with and without the compression stockings. A clear verbal instruction was given to start the jumping. Three output parameters were analyzed to understand the effect of wearing compression stockings on athlete jumping performances. There are knee powers, geometric entropy and ratio of maximum vertical over maximum travelled jumping distance.

IOP Conf. Series: Journal of Physics: Conf. Series **1020** (2018) 012001 doi:10.1088/1742-6596/1020/1/012001

3. Results

a) Knee power
Comparison between the effect of wearing compression stocking on knee power for male and female athlete are shown in Figure 2. Figure 3 shows the comparison effectiveness of wearing compression stocking on mean knee power output for male and female athlete.

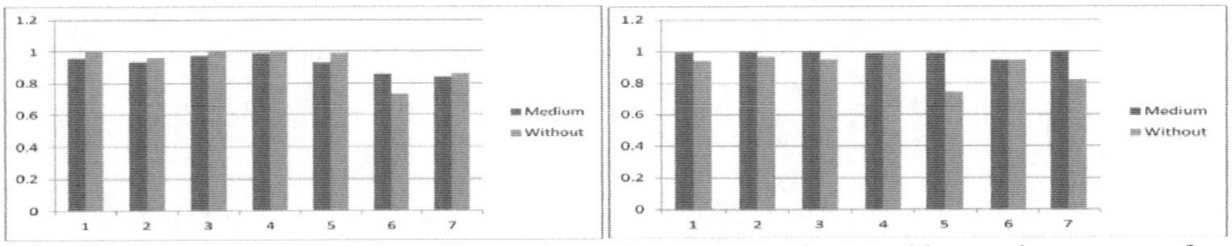

Figure 2: Comparison between the effects of wearing compression stocking on knee power for male and female athlete

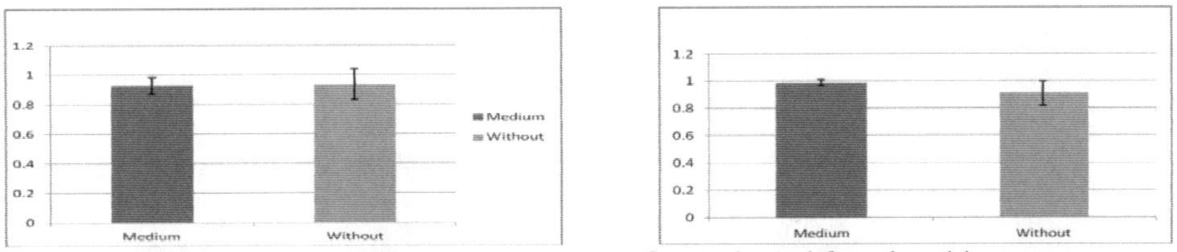

Figure 3: Mean knee power output for male and female athlete

b) Geometric entropy in XY plane
Figure 4 shows the geometric entropy for male and female athlete. Figure 5 shows the mean of the entropy.

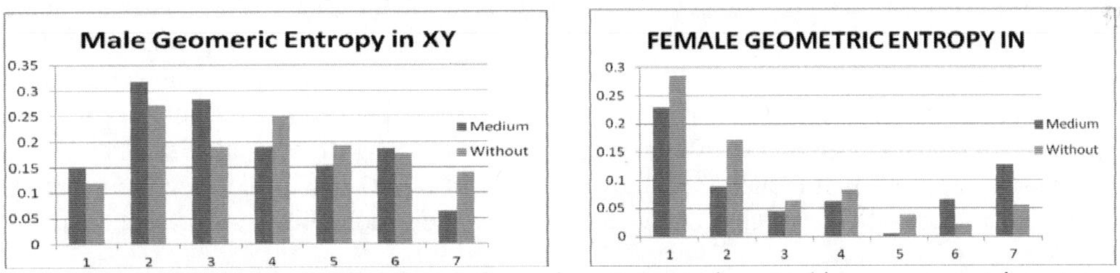

Figure 4: Comparison between the effects of wearing compression stocking on geometric entropy for male and female athlete

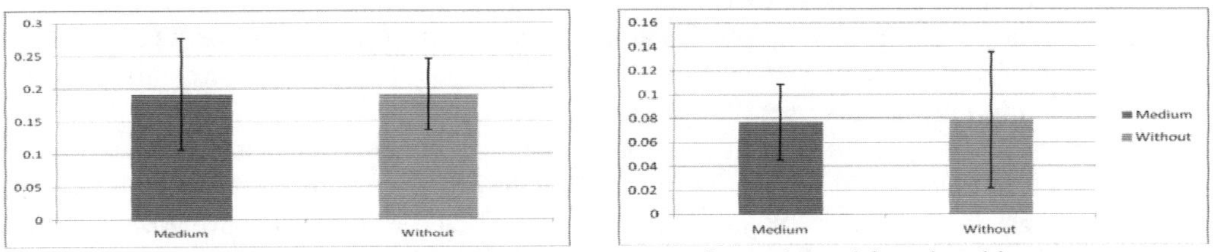

Figure 5: Mean geometric entropy output for male and female athlete

c) Ratio of maximum vertical over maximum travelled jumping distance

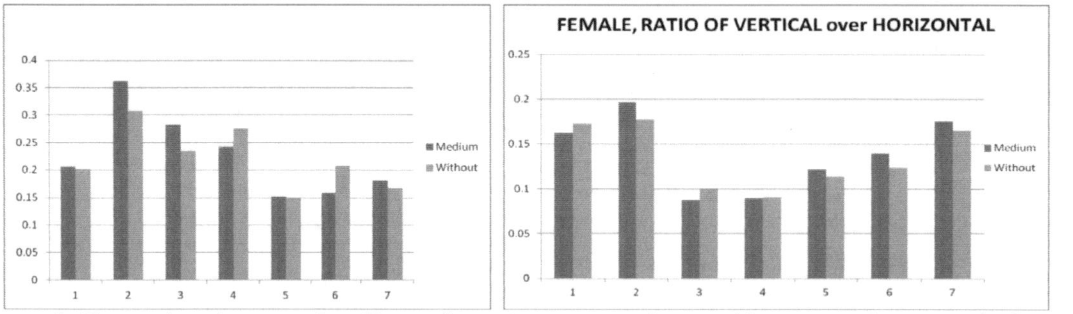

Figure 6: Comparison between the effects of wearing compression stocking on maximum vertical over maximum travelled jumping distance for male and female athlete

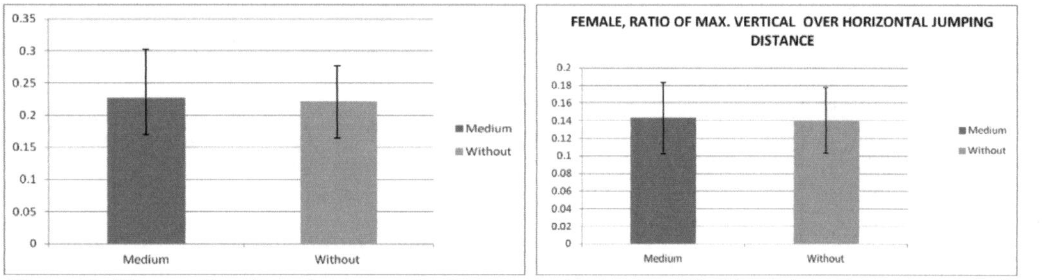

Figure 7: Mean ratio of maximum vertical over maximum travelled jumping distance output for male and female athlete

4. Discussion

There is no significant differences between wearing and not wearing compression stocking ($p<0.05$) on knee power for male athletes. However, for the female athlete, there is a significant difference between wearing and not wearing compression stocking ($p<0.05$) on knee power. The reason is thought to be because for male athlete, the muscle contour is relatively less changes compared to female athlete. Thus, consequently there is less difference between wearing and not wearing compression stocking on knee power of male athlete. However, for female athlete, since by wearing the compression stocking will reduce their muscle movement, thus it will increase their knee power. Perhaps it also relate to the perception of the athlete since their feel more comfortable to jump when their muscle are grip.

Comparing male and female athlete, there is less value of geometric entropy in female athlete than male which shows that less chaotic movement of female athlete shank during jumping. Perhaps, the chaotic movement was influenced by the initial jumping stage momentum where male shows higher initial momentum compare to female case. Although there is a difference value in geometric entropy between genders, however, there is no significant different of geometric entropy between wearing and not wearing compression stocking. The reason is perhaps because the pressure provided by the stocking did not influence the movement of shanks.

Meanwhile, the ratio of maximum vertical over maximum travelled jumping distance in male athlete is relatively higher than female athlete. The reason is thought to be cause by the higher initial jumping momentum and muscle strength in male subjects. In term of wearing and not wearing compression stocking, there is no significant different ($p<0.05$) of ratio of maximum vertical over maximum travelled jumping distance. The reason is perhaps the pressure provided did not influence the ratio much.

IOP Conf. Series: Journal of Physics: Conf. Series **1020** (2018) 012001 doi:10.1088/1742-6596/1020/1/012001

5. Conclusion

This research has studied the effect of wearing customized compression stocking on jumping performance. The stocking was designed and constructed for individual athlete using a developed system and then, experimentally tested. The results show wearing the stocking may help increase female athlete's knee power, but it not affected male athlete as well. Other analysis on geometric entropy and ratio of vertical and horizontal jumping also show no significant difference between wearing and not wearing the stocking.

References

[1] Doan, B., Kwon, Y.H., Newton, R., Shim, J., Popper, E.V.A., Rogers, R., Bolt, L., Robertson, M. and Kraemer, W., 2003. Evaluation of a lower-body compression garment. *Journal of sports sciences*, *21*(8), pp.601-610.

[2] Kraemer, W.J., Bush, J.A., Bauer, J.A., Tripleft-McBride, N.T., Paxton, N.J., Clemson, A., Koziris, L.P., Mangino, L.C., Fry, A.C. and Newton, R.U., 1996. Influence of compression garments on vertical jump performance in NCAA Division I volleyball players. *Journal of Strength and Conditioning Research*, *10*, pp.180-183.

[3] Kraemer, W.J., Bush, J.A., Newton, R.U., Duncan, N.D., Volek, J.S., Denegar, C.R., Canavan, P., Johnston, J., Putukian, M. and Sebastianelli, W.J., 1998. Influence of a compression garment on repetitive power output production before and after different types of muscle fatigue. *Research in Sports Medicine: An International Journal*, *8*(2), pp.163-184.

[4] de Glanville, K.M. and Hamlin, M.J., 2012. Positive effect of lower body compression garments on subsequent 40-kM cycling time trial performance. *The Journal of Strength & Conditioning Research*, *26*(2), pp.480-486.

[5] Scanlan, A.T., Dascombe, B.J., RJ Reaburn, P. and Osborne, M., 2008. The effects of wearing lower-body compression garments during endurance cycling. *International journal of sports physiology and performance*, *3*(4), pp.424-438.

[6] Duffield, R. and Portus, M., 2007. Comparison of three types of full-body compression garments on throwing and repeat-sprint performance in cricket players. *British journal of sports medicine*, *41*(7), pp.409-414.

[7] Wallace, L., Slattery, K. and Coutts, A., 2006. Compression garments: do they influence athletic performance and recovery. *Sports Coach*, *28*(4), pp.38-39.

[8] Salleh, M.N.B., Lazim, H.B.M., Othman, S.N.B. and Merican, A.F.M.B.A., 2013. Development of a flexible customised compression garment pattern design system. *International Journal of Advanced Mechatronic Systems*, *5*(3), pp.202-208.

IOP Conf. Series: Journal of Physics: Conf. Series **1020** (2018) 012002 doi:10.1088/1742-6596/1020/1/012002

Modelling The Effect of Changing Point Systems to Teams' Competition Standing in A Malaysian Soccer Super League

Muhammad Mat Yusof[1], Ruzelan Khalid[1], Mohamad Shukri Abdul Hamid[1], Rosnalini Mansor[1] and Tajularipin Sulaiman[2]

[1]School of Quantitative Sciences, UUM College of Arts and Sciences, Universiti Utara Malaysia, 06010 Sintok, Kedah Darul Aman
[2]Sport Academy, Universiti Putra Malaysia, 43400, Serdang, Selangor Darul Ehsan

mmy@uum.edu.my

Abstract. In a sports league such as in a soccer league, the teams' competition standing is based on a cumulative point system. Typically, the standard point system is given to every single match for win, draw and lose teams is the 3-1-0 point system. In this paper, we explore the effect of changing point systems to teams' competition standing by changing the weightage values for win, draw and lose teams. Three types of point systems are explored in our soccer simulation model; firstly the 3-1-0, secondly the 2-1-0 and thirdly the 4-1-0 point system. Based on the teams participating in a Malaysian soccer Super League, our simulation result shows that there are small changes in term of teams' competition standing when we compared the actual rank and the simulation rank position. However, the 4-1-0 point system recorded the highest Pearson correlation value which is 0.97, followed by the 2-1-0 point system (0.95) and thirdly the 3-1-0 point system (0.94).

1. Introduction

In a sports league such as a soccer league, the cumulative point system is used to figure out the teams' competition standing. The team with the highest cumulative points will become the league champion, followed with the second runner-up team with the second highest cumulative points. The accumulated points are calculated based on the outcomes of individual match, either win, draw or lose. Typically in a soccer league, the standard point system used is the 3-1-0 point system where 3 points for a winning team, 1 point each for draw and 0 points for a losing team. In group stages of soccer World Cup Finals, The Fédération Internationale de Football Association (FIFA) uses the 3-1-0 system. After all match played, the two teams with the highest cumulative points are qualified to the knockout stages. Meanwhile, in soccer leagues in many countries such as in England, Spain, Netherlands, Japan, South Africa, Australia and Malaysia, the 3-1-0 point system is used.

In this paper, we aim to explore the effect of changing point systems to the teams' competition standing. To do this, we have developed a soccer simulation model using a Decision Support System (DSS). This graphical user interface of the soccer simulation model is fundamentally similar with the work in Yusof et. al. (2016) but that work is focusing on the effect of changing various tournament formats such as round-robin and knockout structures. Differently, this paper extended the soccer simulation model in order to study the effect of changing the weightage values for a win, draw and lose team on the teams' competition standing. Here, we have considered three options of point systems; firstly the standard point system used in a soccer league which is 3-1-0, secondly the 2-1-0

Content from this work may be used under the terms of the Creative Commons Attribution 3.0 licence. Any further distribution of this work must maintain attribution to the author(s) and the title of the work, journal citation and DOI.
Published under licence by IOP Publishing Ltd

point system and thirdly, the 4-1-0 point system. We have simulated a league season based on 12 teams participating in a Malaysian Soccer Super League in season 2017.

2. Methodology

In order to investigate the effect of changing point systems to teams' competition standing, we have developed a soccer simulation model. The model is developed using DSS. DSS is basically a computer-based information system supporting decision-making activities to help an organization make relevant decisions. In this paper, we use DSS to assist in modelling the effect of changing point systems without changing the real situation of a Malaysian soccer Super League. The framework for this study is shown in Figure 1.

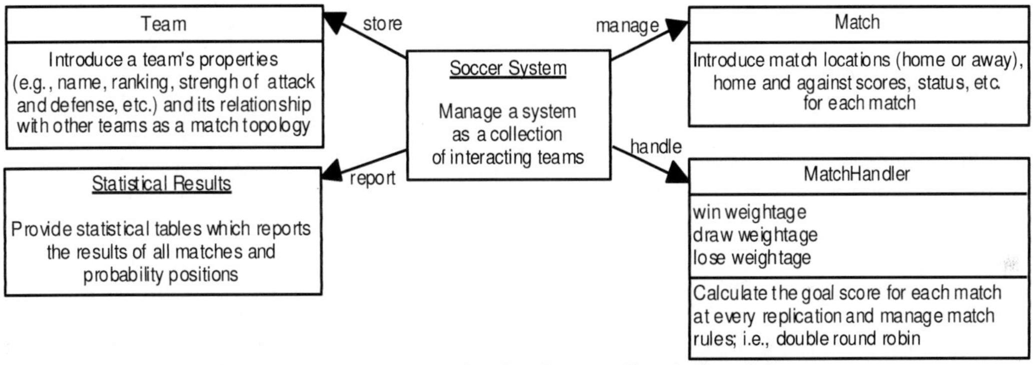

Figure 1. A Framework of A Soccer Simulation System

In our framework, *Team* stores a team's properties; e.g., its name, ranking position and values of attacking strength and defensive weaknesses. *Match* manages match locations, the score and the status (win or lose) of each match. The *Statistical Results* reports the results for all matches and the position ranks of the teams in the league. *MatchHandler* uses the Double Poisson model to calculate the goal scored and conceded for each match at each replication based on the values of attacking strength and defense weakness for each team. This class has properties storing the weightage values of win, draw and lose and the number of replications to be performed to simulate the match outcomes.

The match outcome in our soccer simulation model is based on the number of goals scored and conceded. For simplicity, we used a double Poisson model proposed by Maher (1982). By denoting X_{ij} and Y_{ij} to be the number of goals scored in the match between teams indexed i and j, the model for individual match scores is shown below

$$
\begin{aligned}
(X_{ij}, Y_{ij}) \quad & \sim \text{Poisson}(\alpha_i\ \beta_j\gamma, \alpha_j\beta_i) \text{ for team } i \text{ playing at home} \\
& \sim \text{Poisson}(\alpha_i\ \beta_j, \alpha_j\beta_i\gamma) \text{ for team } j \text{ playing at home}
\end{aligned}
$$

where α_i measures the attacking strength of the teams, the β_j measures the defensive weakness, and γ is allocated for home advantage effects. The team strength and team weakness for 12 teams participated in a Malaysian soccer super league 2017 is shown in Table 1. This strength is calculated based on the 114 individual match outcomes, starting from 20 January 2017 until 5th August 2017.

Table 1. Team Strength and Team Weaknesses (Home advantage: 1.2631)

ID	TEAM	Attacking strength	Defensive Weaknesses
1	JDT	1.42138	1.208459
2	KEDAH	1.65613	1.271639
3	PAHANG	0.91677	1.126739
4	FELDA	1.03439	1.095529
5	SELANGOR	1.6077	1.234631
6	PERAK	1.5785	1.211152
7	PKNS	1.19646	0.965437
8	TTEAM	1.05023	0.994242
9	KELANTAN	1.08488	1.21809
10	MALACCA	1.61216	0.922087
11	SARAWAK	1.43317	1.569173
12	PENANG	1.21892	1.476355

3. Analysis

We develop the soccer simulation model using Visual Basic (VB). The Graphical User Interface (GUI) for this study is shown in Figure 2. As seen in Figure 2, there are several menu options which can be selected such as the structure of the tournament either to choose the round-robin one lag or the round-robin two legs (once played at home and once play at away), the summarization of the report, the types of the point system and finally, the number of the soccer simulation replication.

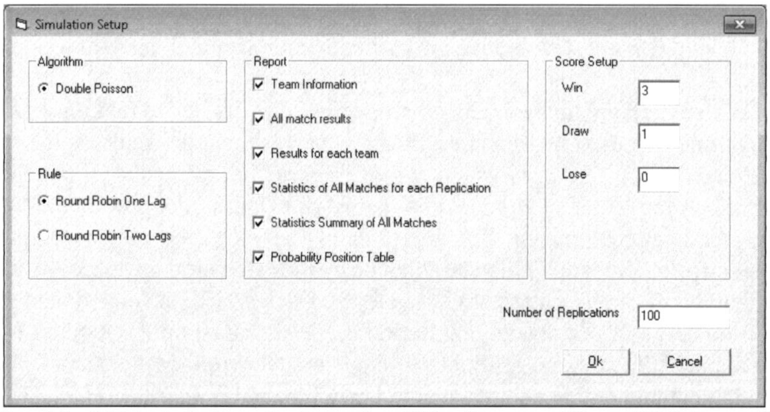

Figure 2. Graphical User Interface for Inputting Win, Draw and Lose Weightage Values

Table 2 shows the example of the outcomes for two-leg round robin, using the 3-1-0 point system and we simulate the soccer league for 100 times. We summarize that after 100 times played in the league, JDT played in total 2200 matches. From all matches, JDT win 792 matches, draw 1340 matches and lose 68 matches. The goal scored and conceded by the JDT team is 1056 and 120 goals respectively. The cumulative points collected by JDT after 100 simulations are 3716 points. Based on the 3-1-0 point system, we find that JDT has the highest probability to become the champion of the league (0.36) compared with other teams.

Table 2. The League Standings Using the 3-1-0 Point System

POS	TEAM	P	W	D	L	GF	GA	GD	PTS	P(win)
1	JDT	2200	792	1340	68	1056	120	936	3716	0.3600
2	KEDAH	2200	636	1413	151	857	245	612	3321	0.2891
3	PERAK	2200	512	1399	289	698	405	293	2935	0.2327
4	FELDA	2200	505	1400	295	683	421	262	2915	0.2295
5	PAHANG	2200	493	1402	305	655	419	236	2881	0.2241
6	SELANGOR	2200	478	1419	303	652	403	249	2853	0.2173
7	TTEAM	2200	308	1383	509	424	676	-252	2307	0.1400
8	PKNS	2200	286	1398	516	406	701	-295	2256	0.1300
9	KELANTAN	2200	267	1437	496	398	687	-289	2238	0.1214
10	MALACCA	2200	200	1431	569	309	788	-479	2031	0.0909
11	SARAWAK	2200	196	1409	595	288	808	-520	1997	0.0891
12	PENANG	2200	142	1339	719	212	965	-753	1765	0.0645

In Table 3, we show the number of league standing position after 100 simulations. The JDT team appears 58 times as a league champion, the Kedah team 25 times and the Pahang team 4 times as league champions.

Table 3. Numbers of Appearance in League Standing Position in 100 simulations.

TEAM	Post 1	Post 2	Post 3	Post 4	Post 5	Post 6	Post 7	Post 8	Post 9	Post 10	Post 11	Post 12
JDT	58	23	12	5	0	2	0	0	0	0	0	0
KEDAH	25	24	17	17	9	3	4	0	1	0	0	0
PAHANG	4	13	9	17	18	19	9	3	5	3	0	0
FELDA	4	15	15	20	13	13	5	8	3	4	0	0
SELANGOR	4	9	18	14	16	15	9	11	2	2	0	0
PERAK	5	11	23	14	18	13	7	4	3	2	0	0
PKNS	0	1	3	2	9	7	13	14	16	13	14	8
TTEAM	0	3	1	7	7	11	8	18	17	16	6	6
KELANTAN	0	1	2	2	5	10	17	14	15	14	11	9
MALACCA	0	0	0	1	3	3	13	12	13	21	18	16
SARAWAK	0	0	0	1	0	2	10	12	18	19	21	17
PENANG	0	0	0	0	2	2	5	4	7	6	30	44

In Table 4, we summarize the probability for each team to become the league champion for various point systems. For example, JDT have the highest probability to win a league using the 3-1-0 point system, followed by the 4-1-0 point system and thirdly by using the 2-1-0 point system.

Based on the Pearson correlation value, we find that there are small changes occurred in the teams' competition standing when we compared the actual rank position and the rank position in our simulation model. All the three point systems we consider give value more than 0.9. Its shows that all the points systems, either 3-1-0, 2-1-0 or 4-1-0 does not affect very much to teams' competition standing.

Table 4. Probability of Teams to Become League Champions and League Standing Position

Actual rank	TEAM	3-1-0 point system		2-1-0 point system		4-1-0 point system	
		Simulation rank	P(win)	Simulation rank	P(win)	Simulation rank	P(win)
1	JDT	1	0.3600	1	0.3134	1	0.3409
2	KEDAH	2	0.2891	2	0.2723	2	0.2841
3	PAHANG	5	0.2241	3	0.2264	4	0.2214
4	FELDA	4	0.2295	6	0.2205	5	0.2195
5	SELANGOR	6	0.2173	5	0.2182	3	0.2386
6	PERAK	3	0.2327	4	0.2327	6	0.2059
7	PKNS	8	0.1300	9	0.1200	8	0.1359
8	T-TEAM	7	0.1400	7	0.1377	7	0.1409
9	KELANTAN	9	0.1214	8	0.1318	9	0.1314
10	MALACCA	10	0.0909	10	0.0991	11	0.0868
11	SARAWAK	11	0.0891	11	0.0882	10	0.0900
12	PENANG	12	0.0645	12	0.0495	12	0.0518
Pearson Correlation Value		0.94		0.95		0.97	

4. Conclusion and Future Works

This paper explores the effect of changing point systems in a soccer league, particularly in a Malaysian soccer Super League. We consider three types of point systems; firstly 3-2-1, secondly 2-1-0 and thirdly 4-1-0 point systems. We simulate the individual match outcome using a double Poisson model and determine the winner, drawer and loser result based on the number of goals scored and conceded. We have developed the soccer simulation model using the Decision Support System.

Based on our analysis, we can conclude that all the point systems we consider give similar interpretation. There are very small changes occurring in the teams' competition standing. JDT and Kedah for example dominate the top two league rank while Malacca, Sarawak, Penang teams dominate the top three in the bottom league. Based on the comparison of the actual teams' competition standing and the simulation model, the result shows that the 4-1-0 point system gives a better result. The Pearson correlation value is 0.97. It shows that there are small changes in term of teams' competition standing when we compared the actual rank and the simulation rank position.

This research however has some limitations and needs further improvement. In our simulation model, we assume that the team strength and team weaknesses for all teams are constant throughout the league. In real situations, every team can make some adjustment in the middle of the league, such as the player's replacement because of the injured, low performance, etc. Secondly, the match prediction we used is very simple, a double Poisson model. Other match prediction models could be explored such as dynamic bivariate Poisson (Koopman and Lit, 2015), a Poisson regression model (Groll et.al, 2015), mixed model effect (McHale and Szcepanski, 2014). Instead of using a round-robin structure, the implication of the teams' competition standing could also be explored for a hybrid tournament and seeding policy rules (Scarf et. al, 2009; Scarf and Yusof, 2011).

Acknowledgement

This work was supported in part by the Ministry of Education Malaysia's under the Research Acculturation Collaborative Effort (RACE) 2013 grant (S/O Code: 12968).

IOP Conf. Series: Journal of Physics: Conf. Series **1020** (2018) 012002 doi:10.1088/1742-6596/1020/1/012002

References

[1] M.M Yusof, R Khalid, T. Sulaiman, M.S.A. Hamid, R. Mansor, N.C. Rosli. E-compare of Soccer Tournament Structures. *International Review of Management and Marketing* **6 (8S),** (2016), 294-298

[2] S. J. Koopman and R. Lit, A dynamic bivariate Poisson model for analyzing and forecasting match results in the English Premier League, *Journal of the Royal Statistical Society: Series A*, **178** (2015), 167-186.

[3] A. Groll, G. Schauberger and G. Tutz, Prediction of major international soccer tournaments based on team-specific regularized Poisson regression: An application to the FIFA World Cup 2014, *Journal of Quantitative Analysis in Sports,* **11** (2015), 97-115.

[4] I. G. McHale and L. Szcepanski, A mixed effects model for identifying goals scoring ability of footballs, *Journal of the Royal Statistical Society: Series A*, **177** (2014), 397-417

[5] P. A. Scarf and M. M. Yusof, A numerical study of tournament structures and seeding policy for the soccer world cup finals, *Statistica Neerlandica*, **65** (2011), 43-57

[6] P. A. Scarf, M. Bilbao and M. M. Yusof, A numerical study of designs for sporting contests, *European Journal of Operational Research*, **198** (2009), 190-198

IOP Conf. Series: Journal of Physics: Conf. Series **1020** (2018) 012003 doi:10.1088/1742-6596/1020/1/012003

A Comparative Study of Data Mining Techniques on Football Match Prediction

Che Mohamad Firdaus Che Mohd Rosli[1], **Mohd Zainuri Saringat**[1], **Nazim Razali**[1], **Aida Mustapha**[1]

[1] Faculty of Computer Science and Information Technology, Universiti Tun Hussein Onn Malaysia, 86400 Parit Raja, Batu Pahat, Johor, Malaysia

E-mail: chemohamadfirdaus94@yahoo.com, zainuri@uthm.edu.my, nazim.iium@gmail.com, aidam@uthm.edu.my

Abstract. Data prediction have become a trend in today's business or organization. This paper is set to predict match outcomes for association football from the perspective of football club managers and coaches. This paper explored different data mining techniques used for predicting the match outcomes where the target class is win, draw and lose. The main objective of this research is to find the most accurate data mining technique that fits the nature of football data. The techniques tested are Decision Trees, Neural Networks, Bayesian Network, and k-Nearest Neighbors. The results from the comparative experiments showed that Decision Trees produced the highest average prediction accuracy in the domain of football match prediction by 99.56%.

1. Introduction

In data generalization, there are two major topics that are concentrated on this subject; Data Mining and Statistics. According to [1], the reason that prediction approaches have been widely used in many companies around the world today is because prediction is the heart of remarkable disciplines in Science. This approach enables companies and organizations to predict and prepare reasonable future plans. One particular applications of prediction is in sport. In order to forecast the future of a particular organization, the logic of prediction is employed. In sport analytic for association football, match prediction is used to predict the matches before it started using a particular technique so that the managers and players able to forecast the possibilities of either win, draw, or loss of the game before it started. Positive outcome will definitely boost up player's spirit but if the outcome is negative, it does not mean to lowered player's sportsmanship but as a guide for them to play extra cautious and provide an appropriate counter-attack.

There are already many researches on the prediction of match outcome in sports, in basketball and football particularly. However, most of those researchers analyzed and predict using a single technique only. There are research that predicts using different techniques but the result might be biased since there are techniques that using a particular technique to increase the accuracy of the prediction. For example, [2] predicted football and basketball using Least Square while [3] predicted football outcome using Neural Tuning and Genetic Tuning fuzzy model. Both researches showed how to match using different techniques with a specific technique create by researchers.

Content from this work may be used under the terms of the Creative Commons Attribution 3.0 licence. Any further distribution of this work must maintain attribution to the author(s) and the title of the work, journal citation and DOI.
Published under licence by IOP Publishing Ltd

To address this gap, a comparative analysis of several data mining techniques using only raw data is essential to find the most accurate techniques using only raw data without any elements that could affect the result like derivation of data or particular technique. This paper is set to achieved three objectives; (1) prediction of football match result (win, lose and draw) using raw data of previous match outcome, (2) model generation of those results using several techniques in data mining, and finally (3) comparative analysis to observe the result of accuracy metrics. This research is limited to previous match's data of three seasons of English Premier League from 2013-2014, 2014-2015, and 2015-2016 totaling 1,140 matches of both home and away.

The rest of the paper is organized as follows: Section II describes the related work on profiling and classification models. Section III presents the method uses in this paper to produce comparative analysis results. Section IV presents the experimental results. Finally, Section V concludes the work and highlights a direction for future research.

2. Related Work

Numerical prediction is a method where a continuous-valued function or ordered-value are predicts by the model constructed. This particular model is also known as predictor. The accuracy of the predictor depends on how well the provided predictor able to predicts the value of predicted attribute for a new data. When compared to classification methodology in data mining, predicted values in prediction methods are usually continuous whereas classifications are discreet, making a classification problem seen as a predictor of classes. Prediction are normally about predicting the future and classification are about classifying the present. Machine learning approaches have been used in predicting match outcome in sports especially for basketball and football association. This prediction is carried out by generating past match results. In this work, four standard prediction algorithms are used for comparative purposes, which are Decision Trees, Neural networks, Bayesian networks, and k-Nearest neighbors.

A decision tree is a structure that includes a leaf nodes, branch nodes, and root nodes. Each leaf node holds a class label, each internal node denotes a test on an attribute, and branch denotes the outcome of a test. The topmost node in the tree is the root node. Decision tree model was used in predicting football matches outcome by [4]. In their experiment, the decision model was developed using the Machine Learning package in order to generate MC4 decision tree. MC4 decision tree is different than the original because it used the tree pruning technique in order to reduce the size and the complexity of the tree. However, MC4 decision tree only managed to predict 45.77% and 41.72% in accuracy for both expert model data and general model data.

Neural Network is a parallel distributed processor that has a propensity for storing experiential knowledge and making it available for users [5]. Neural computing is the study of networks of adaptable nodes which through a process of learning from task. According to [6], neural network model was used to predict the match outcome between two teams during World Cup in 2006. Technically, they proposed a supervised multi-layer perceptron neural network (MLP) with error propagation learning rule (BP) to predict the winning rate for the teams. Then, they compared the result in order to determine the relationship between win and loss. The accuracy for this model was 76.9% excluding the tied games due to limitation in predicting draw outcome.

Belief networks or Bayes nets for short, were the other name for Bayesian Network. It belong to the probabilistic graphical models. These models can be used to represent knowledge about uncertain amount. In particular, the graph contains nodes and edges where nodes represent a random variables and the edges represent the relation of probabilistic dependencies among the corresponding random variables. This Bayesian Network model was used by [7] to predict football outcomes using the Skellam's distribution. This paper focuses on the goal differences instead of the number of goals like most of researchers. The research was made to a specific team with observed number point greater than 1.95 or probability of observed outcome lower than 20%. For instance, the prediction accuracy of win for Bolton team is 53%.

The k-Nearest Neighbors method begins in the early 1950s. The method is labor-intensive when given large training sets, and did not gain popularity until the 1960s when increased computing power became available. Nearest Neighbors are based by comparing a given test tuple with training tuples that are similar to it. [4] used the IB classifier from the family of k-Nearest Neighbors (KNN) in the MLC++ library for predicting matches outcome for Tottenham Hotspur Football Club for 2 seasons. The KNN technique used likeness approach, where a Euclidean distance is measured and plotted on and n-dimensional graph where each dimensional is one of the supplied attributes. KNN technique produced prediction accuracy at 50.58% and 47.21% for two seasons.

3. Methodology

In this section, the dataset, the classification techniques, and the performance measures are presented and discussed. Experiments for Decision Tree, Bayesian Network, and k-Nearest Neighbors are implemented using Weka 3.8 [8] while the Neural Networks algoritm is implemented using the GMDH Shell DS (`https://www.gmdhshell.com/`). In Weka 3.8, the tool uses 10-fold cross-validation method to validate the data. As for classifier, Decision Tree uses J48 classifier while Bayesian Network uses BayesNet classifier. IBk will be the classifier appointed for k-Nearest Neighbors technique. In GMDH Shell DS, the Neural Network technique is implemented using the cross-validation criterion PRR to validate the data.

3.1. Dataset

The dataset chosen to be implement in this research was the data of the English Premier League of 2013 to 2016 season. English Premier League consisted of 20 teams played both home and away totaled 380 matches throughout the season. This is mean there are approximately 1,140 matches for three seasons of English Premier League. The data consists details of the number of full time home and away goal, full time result (win, draw or lose), the number of home and away shots, the number of home and away shots on target and the number of home and away corner. The data was retrieved from the website `http://www.football-data.co.uk`. There are eight features extracted from the datasets, which are `Full Time Home Goal`, `Full Time Away Goal`, `Home Team Shot`, `Away Team Shot`, `Home Shot On Target`, `Away Shot On Target`, `Home Corner`, and `Away Corner` while `Full Time Results` which are win, draw or lose as target class for prediction.

3.2. Evaluation metrics

The performance measures used in this paper is the accuracy. Overall, accuracy measures how often is the classifier or the classification techniques (Decision Trees, Neural Networks, Bayesian Networks, k-Nearest Neighbor) getting correct target class of classification. Equation 1 shows how accuracy is measured.

$$\text{accuracy} = \frac{TP + TN}{P + N} \tag{1}$$

From the equation, TP represents true positives or equivalent to hit, TN represents true negatives or equivalent to correct rejection, P is the total number of real positive cases in the data, and finally N is the total number of real negative cases in the data. In context of football match results, its refer to win, draw or lose target class. For example, if the predicted result of a match hit win but the actual result is lose the result will not be counted in TP and TN. However, it will counted in P which it is actually false positve. Literally, its can be explained that the total number of correct predicted match results (win, draw or lose) are divided with total number of matches.

4. Results

The experiment generated the percentage of accuracy for each technique. All four techniques have been implemented using the specific tool to acquire prediction data. The result were recorded and analyzed. A comparison of the results collected have been made to observe the most accurate technique to predict football match outcome. In order to validate the data, 10-fold cross-validation techniques have been choosen.

4.1. Decision trees

The data retrieved have been imported into Weka 3.8. Then, the data are run using Decision Tree technique. In this case, Weka 3.8 uses Decision Tree J48 Classifier which are also called C4.5 Decision Tree to generate the result. There is no need to discretize the data because C4.5 Decision Tree can handle numeric attributes. The summary of the result is shown in Figure 1. The summary report shows that for 15/16 seasons of English Premier League matches, the accuracy percentage generated was 99.47%.

```
=== Summary ===

Correctly Classified Instances         378               99.4737 %
Incorrectly Classified Instances         2                0.5263 %
Kappa statistic                          0.992
Mean absolute error                      0.0035
Root mean squared error                  0.0549
Relative absolute error                  0.8059 %
Root relative squared error             11.7266 %
Total Number of Instances              380

=== Detailed Accuracy By Class ===

                 TP Rate  FP Rate  Precision  Recall   F-Measure  MCC      ROC Area  PRC Area  Class
                 1.000    0.008    0.983      1.000    0.991      0.988    1.000     1.000     A
                 0.981    0.000    1.000      0.981    0.991      0.987    1.000     1.000     D
                 1.000    0.000    1.000      1.000    1.000      1.000    1.000     1.000     H
Weighted Avg.    0.995    0.002    0.995      0.995    0.995      0.993    1.000     1.000

=== Confusion Matrix ===

   a   b   c   <-- classified as
 116   0   0 |   a = A
   2 105   0 |   b = D
   0   0 157 |   c = H
```

Figure 1. Summary Report of J48 Decision Tree of 15/16 Seasons

4.2. Neural networks

Different tool called the Group Method of Data Handling (GMDH) has been used for running the football prediction using Neural Networks. The same validation techniques have used in order to generate prediction accuracy for Neural Networks technique to be compared to accuracy of Decision Tree, Bayesian Network, and k-Nearest Neighbors, which were all implemented in Weka 3.8. The result of Neural Networks is surprisingly accurate with 100% accuracy throughout 380 matches. Nevertheless, this result does not make Neural Networks as the best technique because the the preduction accuracy is averaged out to take into account performance across three seasons. Figure 2 shows the result for 2015-2016 season.

IOP Conf. Series: Journal of Physics: Conf. Series **1020** (2018) 012003 doi:10.1088/1742-6596/1020/1/012003

Classification accuracy | Table | Raw model

Confusion matrix | ROC curve | Gain chart | Lift chart

Model fit | Predictions

Correctly classified instances 380 100.0% RMSE 0.000 κ-measure 1.000
Incorrectly classified instances 0 0.0% Weighted F-measure 1.000

		Predicted class				
		A	**D**	**H**	**Total**	**Recall**
Actual class	**A**	116	0	0	116	1.000
	D	0	107	0	107	1.000
	H	0	0	157	157	1.000
	Total	116	107	157	380	
	Precision	1.000	1.000	1.000		
	F-measure	1.000	1.000	1.000		
	Baseline	0.695	0.718	0.587	0.413	
	Accuracy	1.000	1.000	1.000	1.000	

Figure 2. Twice-Multilayered Neural Network of 15/16 Seasons

4.3. Bayesian networks

Weka 3.8 was used to generate model using Bayesian Network technique as well as Decision Tree. The data set of English Premier League of 13/14 seasons to 15/16 seasons were then implemented into Weka 3.8 to generate prediction data. The summary report of Bayesian Network of 15/16 Seasons in Figure 3 shows 75.53% of accuracy from 380 matches throughout the seasons.

```
=== Summary ===

Correctly Classified Instances         287               75.5263 %
Incorrectly Classified Instances        93               24.4737 %
Kappa statistic                          0.625
Mean absolute error                      0.2274
Root mean squared error                  0.3229
Relative absolute error                 51.9126 %
Root relative squared error             68.9984 %
Total Number of Instances              380

=== Detailed Accuracy By Class ===

                 TP Rate  FP Rate  Precision  Recall   F-Measure  MCC      ROC Area  PRC Area  Class
                 0.776    0.083    0.804      0.776    0.789      0.700    0.950     0.898     A
                 0.561    0.139    0.612      0.561    0.585      0.433    0.808     0.675     D
                 0.873    0.148    0.806      0.873    0.838      0.718    0.959     0.945     H
Weighted Avg.    0.755    0.126    0.751      0.755    0.752      0.632    0.914     0.855

=== Confusion Matrix ===

  a   b   c   <-- classified as
 90  19   7 |   a = A
 21  60  26 |   b = D
  1  19 137 |   c = H
```

Figure 3. Summary Report of Bayesian Network of 15/16 Seasons

4.4. k*-Nearest neighbors*

The k-Nearest Neighbors technique was also applied to the football dataset by using Weka 3.8. The experiment used IBk classifier with 10-fold cross-validation technique similar to J48 Decision Tree and BayesNet Bayesian Network algorithms. Figure 4 shows the summary report generated by Weka 3.8 of IBk k-Nearest Neighbors technique. The result produced accuracy of 78.16% for 2015-2016 season.

IOP Conf. Series: Journal of Physics: Conf. Series **1020** (2018) 012003 doi:10.1088/1742-6596/1020/1/012003

```
=== Summary ===

Correctly Classified Instances         297               78.1579 %
Incorrectly Classified Instances        83               21.8421 %
Kappa statistic                          0.6687
Mean absolute error                      0.1482
Root mean squared error                  0.38
Relative absolute error                 33.8403 %
Root relative squared error             81.1951 %
Total Number of Instances              380

=== Detailed Accuracy By Class ===

                 TP Rate  FP Rate  Precision  Recall   F-Measure  MCC      ROC Area  PRC Area  Class
                 0.776    0.076    0.818      0.776    0.796      0.711    0.850     0.703     A
                 0.692    0.165    0.622      0.692    0.655      0.511    0.763     0.517     D
                 0.847    0.081    0.881      0.847    0.864      0.771    0.883     0.809     H
Weighted Avg.    0.782    0.103    0.789      0.782    0.784      0.679    0.839     0.695

=== Confusion Matrix ===

   a   b   c   <-- classified as
  90  23   3 |   a = A
  18  74  15 |   b = D
   2  22 133 |   c = H
```

Figure 4. IBk k-Nearest Neighbors of 15/16 Seasons

4.5. Comparative results

The comparative results are shown in Table 1. The results using Neural Network technique generated by GMDH is surprisingly accurate with 100% accuracy throughout 380 matches for season 2014-2015 and 2015-2016. However, the technique still does not suggested as the best since the average accuracy of three seasons been taken into account. The comparative results shown that Desicion Trees have the highest average accuracy by 99.56% follow by Neural Networks and k-Nearest Neighbors technique by 96.83% and 77.54% while Bayesian Networks have the worst average accuracy by 76.41%.

Table 1. Comparison of Accuracy Percentage

Accuracy	Decision Trees	Neural Networks	Bayesian Networks	k-Nearest Neighbors
Season 2013-2014	99.74%	90.50%	77.37%	78.68%
Season 2014-2015	99.47%	100.00%	76.32%	75.79%
Season 2015-2016	99.47%	100.00%	75.53%	78.16%
Average	99.56%	96.83%	76.41%	77.54%

5. Conclusions and Future Works

This study compared four standard data mining algorithms in football match outcome prediction. The experimental results proved that Decision Tree is the most accurate technique to predict football match outcome with 99.56% accuracy as compared to other techniques, which are Neural Network (96.83%), Bayesian Network (76.41%), and k-Nearest Neighbors (77.54%), respectively. However, the research can be improved by studying and creating the same technique for each technique to increase the accuracy of results by using other attributes that correlated to match outcome besides the attributes that have been used in this research such as the distance travelled by away team or current performance of competing teams in the league or the number of yellow or red cards fouled by player which offcourse influece the football match outcome. Moreover, the techiques used in this research is static over time because it depend on past

IOP Conf. Series: Journal of Physics: Conf. Series **1020** (2018) 012003 doi:10.1088/1742-6596/1020/1/012003

match data while the real time football match statistic during match also need to be measured and taking account for prediction (dynamic).

In any sports, knowing the possibilities of winning a game is essential. Result of outcome prediction can be used to elevate player's sportsmanship or to prepare a counter-attack against their opponents. Match outcome prediction should become a trend in our national sports in order to produce a profitable and quality organization of a particular sports while increases athletics' performances. Management of a particular sports also can used this prediction method to forecast the future of an organization.

Acknowledgements
This project is sponsored by the Malaysian Ministry of Higher Education (MOHE) under the Research Acculturation Grant Scheme vot 1513.

References
[1] Dhar V 2013 *Communications of the ACM* **56** 64–73
[2] Stefani R T 1977 *IEEE Transactions on systems, man, and cybernetics* **7** 117–121
[3] Rotshtein A P, Posner M and Rakityanskaya A 2005 *Cybernetics and Systems Analysis* **41** 619–630
[4] Joseph A, Fenton N E and Neil M 2006 *Knowledge-Based Systems* **19** 544–553
[5] Rumelhart D E, Hinton G E and Williams R J 1988 *Cognitive modeling* **5** 1
[6] Huang K Y and Chang W L 2010 A neural network method for prediction of 2006 world cup football game *Neural Networks (IJCNN), The 2010 International Joint Conference on* (IEEE) pp 1–8
[7] Karlis D and Ntzoufras I 2009 *IMA Journal of Management Mathematics* **20** 133–145
[8] Witten I H, Frank E, Hall M A and Pal C J 2016 *Data Mining: Practical machine learning tools and techniques* (Morgan Kaufmann)

IOP Conf. Series: Journal of Physics: Conf. Series **1020** (2018) 012004 doi:10.1088/1742-6596/1020/1/012004

A Review on Football Match Outcome Prediction using Bayesian Networks

Nazim Razali[1], **Aida Mustapha**[1], **Sunariya Utama**[2], **Roshidi Din**[2]

[1] Faculty of Computer Science and Information Technology, Universiti Tun Hussein Onn Malaysia, 86400 Parit Raja, Batu Pahat, Johor, Malaysia

[2] School of Computing, College of Arts and Sciences, Universiti Utara Malaysia, 06010 Sintok, Kedah, Malaysia

E-mail: nazim.iium@gmail.com, aidam@uthm.edu.my, sunariya.utama@gmail.com, roshidi@uum.edu.my

Abstract. Modelling association football or soccer prediction model has become a popular issue among scholars in the last few years. This paper is set to study a Bayesian networks (BNs) for football match outcome prediction. The main objective of this paper is to review the application of Bayesian networks (BNs) in football match prediction. This review may be useful for future researchers in developing new prediction model using BNs that incorporate several new features alongside with other existing features that contribute to increase the predictive performance of football prediction model.

1. Introduction

Association football (British English) or soccer (American English) is one of the most popular sports in the world. People of the world enjoy playing the sport or watching the matches played by amateur until professional players. The analytics for association football comprises into two parts. First, performance analysis can be classified according to two aspects, in terms of individual performance or team performance. Second is match prediction, which can also be classified into two perspectives, in terms of business purposes, generating profits for betting and bookmakers. The other part is benchmark for evaluating the chance of win, draw or lost for the team before the match begins. This help the managers and players with fresh insight in preparing the countermeasures such as using defensive mentality and offside trap if the opponents is stronger. Figure 1 shows that the categories of sports analytis.

The performance analysis is invaluable in discovering the principal of the observed outcome; win a match or score goals. Based on network theory analysis of football strategies, some of the teams participating in 2010 World Cup are analyzed in terms of team passes. The result produced from [1] were the closeness (how well connected are a player), betwenness (how dependent team to the player), page rank (how important are a player), and clustering. The limitation of this analysis is that it did not consider defensive strength of the opponents. It also did not focus on passing accuracy and shot analysis, which are fundamental in a football match. Otherwise, the match prediction obtains the profits from betting will not be touched in this research because Islam does not tolerate with gambling as Islam strictly prohibit its followers from gambling.

Content from this work may be used under the terms of the Creative Commons Attribution 3.0 licence. Any further distribution of this work must maintain attribution to the author(s) and the title of the work, journal citation and DOI.
Published under licence by IOP Publishing Ltd

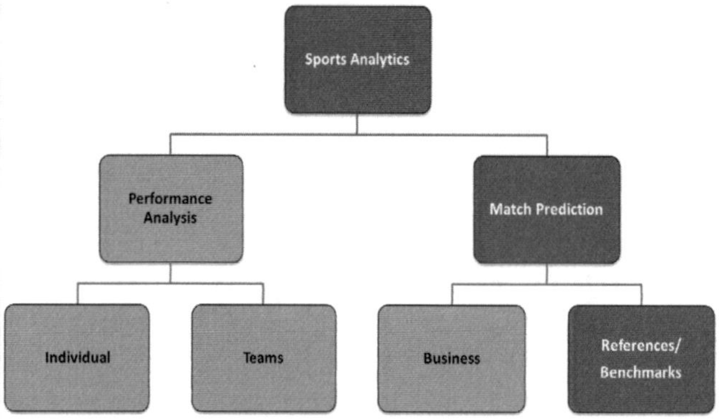

Figure 1. Classification of sport analytics categories

Prediction in football has become an intriguing research problem, partly because of the difficulty, because there are many factors can influence the football matches outcome, such as teamwork, skills, weather, home advantage and many others. The challenge is also faced by the experts in football as well because it is very difficult to anticipate the actual results of football matches. Anything can be happened within 90 minutes of football match with none or extra time such as injury or sent off of player by red card. Luck also can be a factor that influence the results of football matches as the strong team are not necessarily win the match against the weak team. Due to the different factors that influence the football match, some research resort to Bayesian Networks (BNs) due its proven applicability in predicting the weather, sports, disease, technology and etc.

Bayesian Networks have been widely used in researches or studies in field of Artificial Intelligent (AI) and Statistics. The Artificial Intelligence and Statistics shared the same goal of modelling real-world phenomena and what the slightly difference between them is AI researches give priority to knowledge based approach for achieving objective study while statisticians give priority to data [2]. The remainder of this paper is organized as follows. Section II introduces the theoretical foundation of Bayesian Networks. Section III reviews related work in football prediction that applied Bayesian Networks. Section IV discusses the different implementation of BNs in match outcome prediction and finally Section V concludes with some indication of future works.

2. Bayesian Networks (BNs)

Bayesian networks (BNs) was known in many names such as belief networks, Bayes nets, probabilistic belief, Bayes network and causal network are actually belonging to the family of probabilistic graphical models. It is used to model probabilistic relations among random variables which mean the items involved vary in some random or unexplained manner [3]. [4] points out that BNs is needed to encounter the knowledge where the inferences drawn are subjective, uncertain and incomplete in the beginning reasoning process for probability theory. The probability theory is very useful in order to calculate and permits inferences to flow in two ways; from hypothesis to evidence (predictive) as well as from evidence to hypothesis (diagnostic). It showed that BNs can be used to predict the outcomes or diagnose causal effect (if structures are known).

In addition, BNs can be used to discover causal relationship (if structures are not known). Therefore, BN offered solution for inferential reasoning, decision and uncertainty using quantitative and qualitative components. The qualitative component is represented by the network structure while the quantitative component is represented by the network parameters, which are

IOP Conf. Series: Journal of Physics: Conf. Series **1020** (2018) 012004 doi:10.1088/1742-6596/1020/1/012004

the conditional probability distributions of the nodes in the network. The main idea of Bayesian network is actually come from the work of Thomas Bayes called the Bayes' theorem. The Bayes' theorem is represented in Equation 1 or Equation 2.

$$P(\text{cause}|\text{effect}) = \frac{P(\text{effect}|\text{cause}).P(\text{cause})}{P(\text{cause})} \quad (1)$$

where:

- $P(\text{cause}|\text{effect})$ is the conditional probability quantifies the relationship in causal direction.
- $P(\text{effect}|\text{cause})$ describes the diagnostic direction.
- $P(\text{effect})$ as the evidence of the unknown cause.
- $P(\text{cause})$ as the unknown variable that needed to be determined.

$$\mathrm{P(X|Y)} = \frac{\mathrm{P(Y|X).P(X)}}{\mathrm{P(Y)}} \quad (2)$$

where:

- $P(X)$ is the prior probability or marginal probability of X.
- $P(X|Y)$ is the posterior probability or conditional probability of X given Y.
- $P(Y|X)$ is the conditional probability of Y given X (the likelihood of data Y).
- $P(Y)$ is the prior probability or marginal probability of data Y (the evidence).

Equation 1 and Equation 2 explained the Bayes' theorem both literally and technically. Upon observation of evidence, the prior probability of a hypothesis changes to a posterior probability. This can be illustrated in the following way; before observing the data $Y/effect$, the distribution for $P(X)/cause$ would be known as the prior distribution because $X/cause$ does not have any information about $Y/effect$. This distribution quantifies the parameter values before $Y/effect$ is observed. However, once the data $Y/effect$ has been observed, this prior distribution is updated to the posterior distribution $P(X|Y)/P(\text{cause}|\text{effect})$. The term "posterior" means that the probability is derived from data $Y/effect$. The task of computing this probability distribution for a set of query nodes, given the values for some evidence or observation nodes is how we reason with probabilistic systems like BNs or dynamic BNs.

2.1. *Qualitative Component of Bayesian Networks*

A qualitative component of Bayesian network (BN) is represented the network structure called directed acyclic graph (DAG). A directed acyclic graph implemented set of nodes that characterize random variables from the domain and directed edges connecting nodes to represent the conditional dependencies between nodes. Though, the directed edges unable on the form of any directed cycles. The Bayesian network in form of directed acyclic graph is stated in Figure 2 as follows.A qualitative component of Bayesian network (BN) is represented the network structure called directed acyclic graph (DAG). A directed acyclic graph implemented set of nodes that characterize random variables from the domain and directed edges connecting nodes to represent the conditional dependencies between nodes. Though, the directed edges unable on the form of any directed cycles. The Bayesian network in form of directed acyclic graph is stated in Figure 2 as follows.

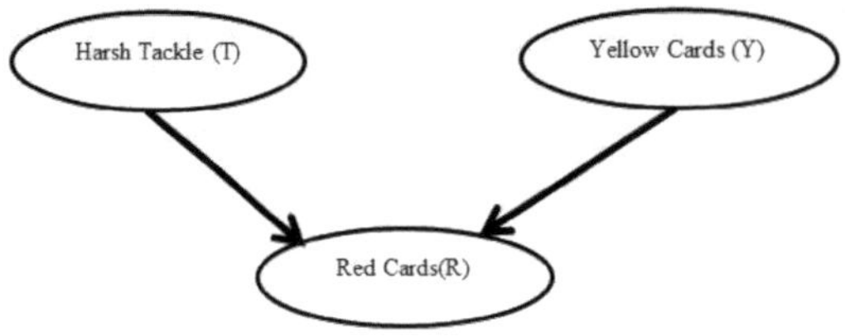

Figure 2. A simple Bayesian network in Directed Acyclic Graph

In Figure 2 showed there are two events can cause red card delivered by referee to the footballs player which are either due to harsh tackle or the second yellow card according on how the referee interpret the situation and judgement. Therefore, all three variables have two possibility which is $true(T)$ and $false(F)$.

2.2. Quantitative Component of Bayesian Networks
A quantitative component of Bayesian network is used to encode the local influences among the random variables using probability distributions [5]. Thus, it is actually can be explained that a quantitative component of Bayesian network is in form of a set conditional probabilities for discrete data or probability density function for continuous data. The BN in form of conditional tables is stated in Figure 3.

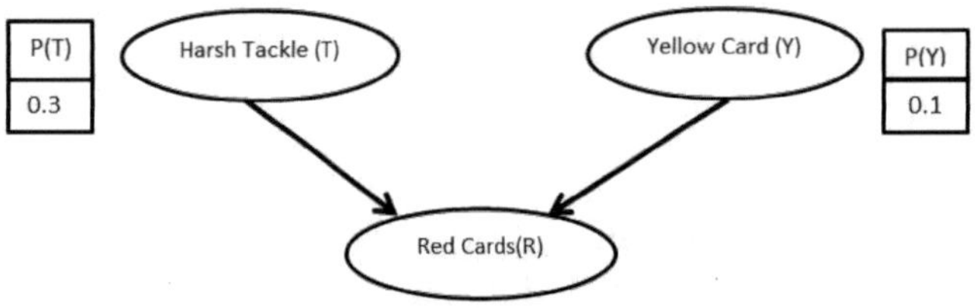

P(T)	P(Y)	P(R)
F	F	0.02
F	T	0.2
T	F	0.5
T	T	0.97

Figure 3. A simple Bayesian network with conditional probability tables (CPT)

Based on Figure 3 presented the probabilities of two events; harsh tackle and yellow card affect the red card. Given the harsh tackle also give possibility for the player receiving yellow card or red card. It explained that each node in a Bayesian network must has a probability distribution table, where the probability distribution of the node is given respected to its parents, the equation shown as stated in Equation 3:

$$P(X_i|Parents(X_i)) \tag{3}$$

Parents (X_i) means the all the nodes that have an arrow to node X_i. For example, yellow card and harsh tackle is a parent of red card because harsh tackle and yellow card can cause red card or red card is a consequence of yellow card and harsh tackle. By multiplying all nodes given their parents, the joint distribution of the network is found as stated in Equation 4:

$$P(X_1, X_2, ..., X_n = \prod_{i=1}^{n} P(X_i|Parents(X_i)) \quad (4)$$

Where application of the chain rule of probability theory shown in Equation 3 allowed factorization of joint probabilities as in Equation 4,

$$\begin{aligned} P(X_1, X_2, ..., X_n &= P(X_1) \times P(X_2|X_1), ..., \times P(X_n|X_1, ..., X_{n-1}) \\ &= \prod_{i=1}^{n} P(X_i|Parents(X_i)) \end{aligned} \quad (5)$$

Given probability 0.3 for $P(T)$ means that it is 30% chance of tackle committed by player considered harsh but there is 70% chance the tackle committed is considered legal. Furthermore, 97% chance of receiving red card when harsh tackle was committed and the player already received yellow card once. It is because the nodes representing a red card are conditioned on the nodes for harsh tackle and yellow card. Its probabilities are given respected to its parents nodes. Therefore, only 3% chance that the player will not gained red card. This might be the result when the referee making wrong decision or there are some other events that might influence the judgement of referee such as the referee failed to interpret the situation.

3. Application of Bayesian Networks in Football Prediction

Application of Bayesian networks (BNs) in football prediction is all started by [6] that improvised the Poisson distribution of [7] and [8] for predicting football match outcome using dynamic Bayesian networks (DBNs). [7] present the strength of defensive and offensive strength of home and away team with goal scoring distribution using independence Poisson distribution. However, Maher model just an ex-ante model which means use actual result for expressing the defensive and offensive of team and cannot predict the future. [8] enhanced the Maher model by implementation of bivariate Poisson distribution for football prediction. Next, [6] applied the concept of [8] in Bayesian way where it can incorporate more uncertainty or time-variation parameter such as psychological effect.

[9] proposed an expert Bayesian Networks (BNs) model and compared with four alternative machine learning (ML) technique, which are naive BNs, data driven BNs, K-Nearest Neighbor and Decision Tree. The efforts of the study was restricted to all league matches by Tottenham Hotspur F.C. during seasons 1995/1996 and 1996/1997. The features used in expert BNs was presence of three players, the position of a player who always play, quality of opposing team in 3-point scale, venue, attacking force quality, overall team quality, and performance given their own and opposition quality. The experiment was repetitive with various machine learning approaches using the same set of features.

[10] proposed framework called FRES (Football Result Sports System) for Bayesian inference and rule-based reasoning. There are twenty seven features used in this model, which were sourced from match statistics. The features were used to construct four Bayesian networks for offense, defense, possession, and fatigue. It is found that the produced a Relative Rank Error of 5.920 and Root Mean Square Error of 5.306 during forecasting the result of World Cup 2002. The model considered dynamically in order to changes team strategies throughout the entire match, which in turn influenced the prediction results.

Then, [11] proposed a Bayesian hierarchical model with aim to guesstimate the appearances that bring a team win and lose a match, predict the outcome of a particular match and test the models predictive strength. The model test used the data Italian Series A Championship 2007-2008. However, the model have drawback called over shrinkage. The over shrinkage occurred when the grand mean of the overall observation has been pulled by some of the extreme occurrences such as there are few teams that have very good performance and few teams that struggling for relegation due to ineffective performance faced each other in a match. It is to changeling in order to estimate and capture each team performance because the strong team may overestimate the weak team.

Other Bayesian network model for forecasting match was is presented in [12]. The data used for training were the English Premier League season 1993/1994 to 2009/2010. The two types of features used the objective information and subjective information. The objective information is the total points previous five season and total points of current season. Whereas the subjective information are the expert belief of team strength, key player and first team player availability, previous form, motivation, team spirit, managerial issue, head-to-head bias, days since last match, first team player rested during last match, previous match toughness, and national team participant. The accuracy of the model was evaluated using Rank Probability Score (RPS) on forecast that make use objective information only and forecast that make use of objective and subjective information. Based on the forecast of English Premier League season 2010/2011 matches, objective and subjective information forecast possess better accuracy than objective information only forecast. Figure 4 summarized the dataset used for prediction model using Bayesian approaches.

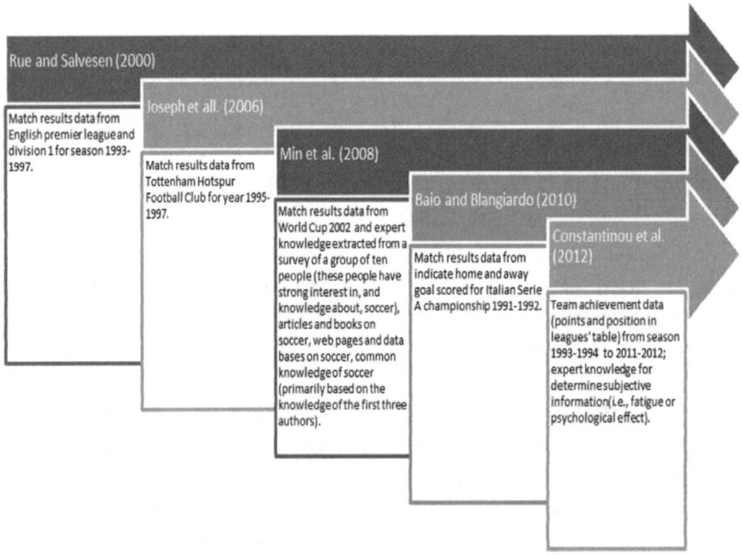

Figure 4. The dataset used by previous researchers for BNs football prediction model

Based on Figure 4, it is discovered that the datasets used for implementing the football prediction using Bayesian approaches generally from match statistical data or team statistical data and even from expert judgment of the club for football league in the several season.

4. Discussion

The reviews have shown that existing models rely heavily on historical data for team features (e.g. total number of shots, the number of goals scored or conceded by a team in match and etc.), matches features (e.g. results of previous meeting between team 1 and team 2, results of past matches for team 1 and etc.) and team achievement (ranking of a team in league or world, total number of points gained by team during or end of the season and etc.) which are mostly readily accessible via FIFA Association website and news broadcasting. However, some of them used expert knowledge or specific datasets which they need to get the help or service from football expertise or need to buy from sports data provider such as Opta (`http://www.optasports.com/`) and STATS (`https://www.stats.com/`) in order to increase their prediction model accuracy. Thus, this expert type of data gained from expertise sometimes can be biased toward prediction model itself because every expert has different view and opinion which are subjective and may affect the data reliability. Besides, sports data gained from data provider can be expensive if a lot of data was needed for modelling prediction model (learning state).

[13] point out that harvesting more match data can generate even richer prediction model for football. He improvised the works of [6] by adding the player statistic data, the total number of shots fired and shots on target which outperformed other traditional statistical football prediction model [7, 8, 6] which based on the team past results data. As for football sports, the use of Bayesian approaches in predicting match outcomes are gaining momentum due its capability for predicting uncertainty and reasoning over time. This is particularly important considering the dynamic nature of football matches within 90 minutes. The pi-rating system [12], which was developed based on Bayesian approach outperformed the Elo rating model [14] and its improved version in [15] because the rating models were becoming more dependent towards the bookmaker odds for prediction. As the prediction model improved time by time, they become more complex in how they treat data in order to incorporate more features or variables and associate dynamically with time like the work of [10]

In summary, the research in match outcome prediction for football thrives in two major directions; the statistical approach and the Bayesian approach. Contributions in machine learning approach are expected to grow further in the future as the fact that sports analytics in football requires more features from various perspectives such as the players' strength and team abilities on top of the past match statistics. The performance of existing machine learning approach is also fair as shown by the work of [16] (Fuzzy model) and [17] (Neural Network model), whereby the prediction accuracy for both machine learning techniques is reasonably good for both international and domestic football competition. However, [9] showed that the Bayesian approach outperformed the k-nearest neighbor and decision tree algorithms under the machine learning approach.

This review has set two potential avenues for future work on the topic of football match prediction. The first is to model the rich data based model which may enhance the cabilities of predictive accuracy. The second is to make the prediction models to be more dynamic and sensitive to real-time changes during the match in order to cater uncertain events such as injury, fouls, and substitutions (team as well as opponents). For example, a match prediction model should be able to put face values on each individual player at specific point of time on an off the match.

5. Conclusions

According to FIFA Big Count in 2016 through FIFA official website (`http://www.fifa.com/worldfootball/bigcount/allplayers.html`), Malaysia as of 3 May 2016 has 585,730 players in which 9,930 are registered and the remaining 575,800 are unregistered players. This makes less than one percent from the 265,000,000 football players across the world. Malaysia also hosts 110 clubs and a total of 11,810 officials in football matches. The findings of the study will contribute greatly on sports development especially in association football or soccer in our country with need a touch in order to improve our teams ranking, pleasing our football fans and shut the criticize towards our football national team. The analysis provided is hoped to help coaches to evaluate and estimate their players performance, hence a more accurate network structure between players and the match outcomes (win, draw and loss). As the results, this findings are invaluable for developing and forming a team with the highest winning probability.

From the economic perspective, the improvement of club performance on the field would lead to improved revenue from stadium attendance and merchandise sold, thus lead to national economic growth. Besides, a new BNs prediction model can be used to set benchmark which may assists the managers and players acts with more cautions and prepare the alternative way to beat the opponent that stronger than them such as setting up offside trap, implement defensive mentality and play counter attack.

Acknowledgements
This research is funded by the Ministry of Higher Education Malaysia (MOHE) under the Research Acculturation Collaborative Effort (RACE) Grant Scheme Vot 1513.

References

[1] Pena J and Touchette H 2012 *Sports Physics: Proc. 2012 Euromech Physics of Sports Conference* 517–528

[2] Heckerman D, Geiger D and Chickering D 1995 *Machine Learning* **20** 197–243

[3] Heaton J 2013 *Forecasting & Futurism 7* 6–10

[4] Pearl J 1985 Bayesian Networks: A Model of Self-Activated Memory for Evidential Reasoning (No: CSD-850021)

[5] Kersting K and Raedt L 2000 Bayesian Logic Programs (Report No. 151)

[6] Rue H and Salvesen O 2000 *Journal of the Royal Statistical Society: Series D (The Statistician)* **49** 399–418

[7] Maher M 1982 *Statistica Neerlandica* **36** 109–118

[8] Dixon M J and Coles S G 1997 *Journal of the Royal Statistical Society. Series C: Applied Statistics* **46** 265–280

[9] Joseph A, Fenton N E and Neil M 2006 *Knowledge-Based Systems* **19** 544–553

[10] Min B, Kim J, Choe C, Eom H and (Bob) McKay R I 2008 *Knowledge-Based Systems* **21** 551–562

[11] Baio G and Blangiardo M 2010 *Journal of Applied Statistics* 1–13

[12] Constantinou A C, Fenton N E and Neil M 2012 *Knowledge-Based Systems* **36** 322–339

[13] Langseth H 2013 *Frontiers in A.I. and Applications* **257** 165–174

[14] Hvattum L M and Arntzen H 2010 *International Journal of Forecasting* **26** 460–470

[15] Leitner C, Zeileis A and Hornik K 2010 *International Journal of Forecasting* **26** 471–481

[16] Rotshtein A P, Posner M and Rakityanskaya A B 2005 *Cybernetics and Systems Analysis* **41** 619–630

[17] Huang K y, Member S and Chang W l 2010 *The 2010 International Joint Conference on Neural Networks (IJCNN)* 18–23

IOP Conf. Series: Journal of Physics: Conf. Series **1020** (2018) 012005 doi:10.1088/1742-6596/1020/1/012005

P-Soccer: Soccer Games Application using Kinect

Mohamad Fahim Mohamed Nasir[1], Suriawati Suparjoh[1], Nazim Razali[1], Aida Mustapha[1]

[1] Faculty of Computer Science and Information Technology, Universiti Tun Hussein Onn Malaysia, 86400 Parit Raja, Batu Pahat, Johor, Malaysia

E-mail: ai130156@siswa.uthm.edu.my, suriati@uthm.edu.my, nazim.iium@gmail.com, aidam@uthm.edu.my

Abstract. This paper presents a soccer game application called P-Soccer that uses Kinect as the interaction medium between users and the game characters. P-Soccer focuses on training penalty kicks with one character who is taking the kick. This game is developed based on the Game Development Life Cycle (GDLC) methodology. Results for alpha and beta testing showed that the target users are satisfied with overall game design and theme as well as the interactivity with the main character in the game.

1. Introduction

Soccer or association football is among the well-known sports in Malaysia and the world. Soccerl is a team sport that is currently at rise and popular in Southeast Asia continent. There are various types of football tournaments which witnessed by people around the world such as the Barclays Premier League (England), the Spanish League, World Cup and etc. Eleven individual players are required to complete a football team on the field. As for penalties, only one individual player involved for each team. A forward may take the penalty while a goalkeeper will protect the goal from be conceded.

Nowadays there are various forms of football games applications that have been developed to assist players performing training either computer or mobile phone platform based. The Penalty Shoot-out and Soccer 2015 Fast Reflex is among existing gaming applications that already be developed. Therefore, the P-SOCCER gaming application that uses the Kinect technology was developed to assist players in training. Kinect is a technology where the device provides a natural user interface for the users to interact with the application without physical contact to any device [4]. This technology has introduced a new experience to the application users with a more flexible and intuitive interaction. In the proposed P-Soccer games application, Kinect is integrated with the game to allow game players to interact with Kinect while training the penalty kicks.

Based on observations of some existing applications, most applications use different type of platforms and also different type of interaction between users and applications. Kinect technology is a technology that introduce a new mode of interaction that allow the users to use their body movement. The objective of this project is to design and develop a soccer game application based on strategic module of football training. The proposed application is then integrated with the Kinect technology to provide a more intuitive and interactive behavior in game.

Content from this work may be used under the terms of the Creative Commons Attribution 3.0 licence. Any further distribution of this work must maintain attribution to the author(s) and the title of the work, journal citation and DOI.
Published under licence by IOP Publishing Ltd

IOP Conf. Series: Journal of Physics: Conf. Series **1020** (2018) 012005 doi:10.1088/1742-6596/1020/1/012005

The remaining of this paper proceeds as follows. Section II presents existing games most similar to the proposed P-Soccer, Section III presents the methodology, covering from eliciting user requirements, game design, storyboard, and implementation. Section IV reports on the testing and finally Section v concludes with some planning for future works.

2. Related Work

The earliest games application is called 'Pong' the Atari Inc. in 1972 [6]. There are many genres in games development including the action genre, informational games, sports, educational and many others. In sports, two existing games most similar to the proposed P-Soccer are the Fast Reflex Soccer and Penalty Shootout. In Fast Reflex Soccer, the game is in the form of 2D on mobile platform. Users can choose whether to enter single play or dual play. The module 'Practice' allows users to train before the actual game. The gameplay shows a gray circle and users have to touch the screen to begin kick. The user has to score the goal when the circle changes color from grey to red. Figure 1 shows the interface for Fast Reflex Soccer games.

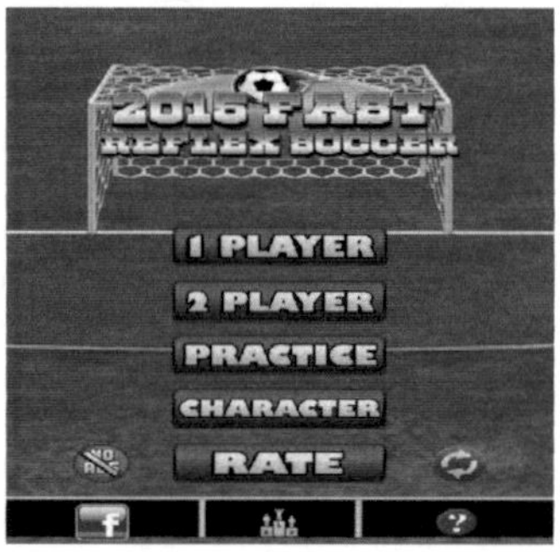

Figure 1. Fast Reflex Soccer

In Penalty Shootout, users have to choose a team and are given five attempts to take the penalty. The user with the highest score will get the highest mark and win the round. To kick the ball, user has to choose the direction by calibrating the angle. Figure 2 shows the interface for Penalty Shootout games.

Figure 2. Penalty Shootout

Table 1 shows the feature comparison between the proposed P-Soccer application against existing games. The comparison takes into account the interface dimension, number of modules, interaction modes, and application platform.

Table 1. Comparison of Application Technologies

	Penalty Shootout	Fast Reflex Soccer	P-Soccer
Dimension	2D	2D	3D
Module	Three modules: • Play • Highscore • More Games	Three modules: • Choose Player • Practice • Choose Character	Two modules: • Play • Info
Interaction Mode	Mouse	Touchscreen	Kinect Device
Platfrom	Desktop	Mobile Smartphone	Desktop

3. Methodology

The methodology adopted to develop the proposed P-Soccer games is the Game Development Life Cycle (GDLC). GDLC consists of five phases, which are initiation, pre-production, production, testing, and the release.

- Initiation: Initiation stage begins with seeding ideas. Among the early decisions include identifying common problems in soccer training, the development platform, target users and technologies to support the game development.
- Pre-production: Pre-production involves the game design including the application genre, characters, and the types of user interaction that will be supported. In P-Soccer, the genre is games and there is only one character involved, who is the player taking the penalty kick. Aside the main character, other important element in the design is the real environment, which includes the field, goalposts, and spectators in the stadium.
- Production: Production is the most important phase in GDLC methodology, which involves production of game assets, source code, and integration between the two elements. The game is then connected to the Kinect device.
- Testing: Testing involves two phases, which are alpha testing and beta testing. Alpha testing is set to evaluate the games itself, which includes functionality testing, device integration, and character matching.
- Release: Deployment of the P-Soccer games. All assets and models are integrated using the Unity 3D software.

3.1. User requirements

The target users for P-Soccer are the soccer varsity players for Universiti Tun Hussein Onn Malaysia (UTHM). System requirements are elicited from the coach, who also identified the target users among the midfileders and attackers. According to the coach, the most important aspect in taking penalty kicks is the accuracy of the kick. To emphasize this importance, P-Soccer provides three views in the game; (1) player position, (2) target, and (3) the kick itself. The main character in P-Soccer responds to the movement of the game player based on interactivity provided by the Kinect technology.

3.2. Game design

Graphical design is important in game development. The main objective of design in P-Soccer is interactivity and intuitivity. The design for models are also in accordance to the metaphors and environment based on the application theme and suitable for target users. P-Soccer, was

IOP Conf. Series: Journal of Physics: Conf. Series **1020** (2018) 012005 doi:10.1088/1742-6596/1020/1/012005

developed using the Kinect device v.1, which is primarily used for tracking user movement when playing this games. The software rused was the Kinect Studio: Kinect Developer Toolkit v1.7. This software is important to track and display real environment via the Kinect device camera.

Other softwares used to develop P-Soccer include Adobe Flash Professional and Adobe Photoshop CS6 for interface design, Audacity for audio, Autodesk 3Ds Max 2016 for model design, and finally Unity 3D version 5.2.1 as the development platform. Figure 3 shows the interactive button for the main modules in P-Soccer, which are `Play` and `Info`.

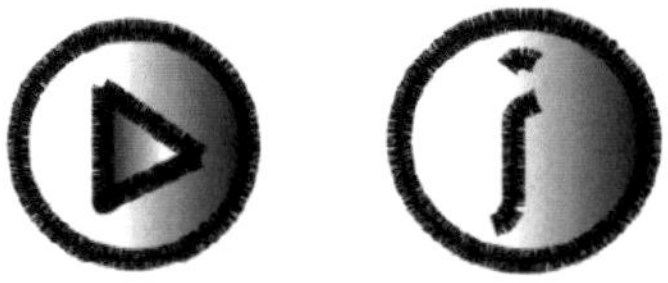

Figure 3. Buttons in P-Soccer

Figure 4 shows joint matching among characters in P-Soccer.

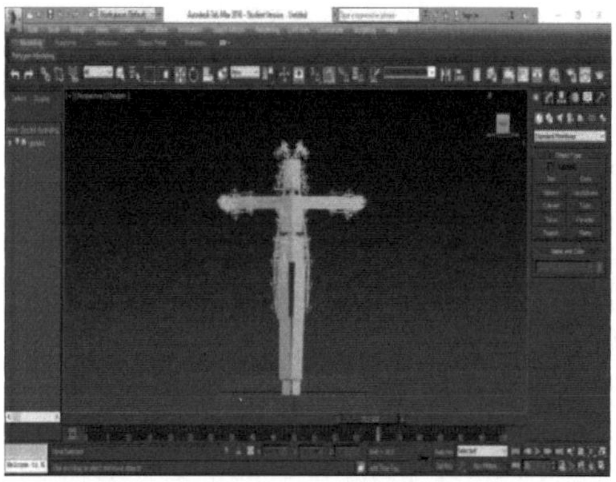

Figure 4. Characters in P-Soccer

3.3. Storyboard

In the storyboard, there are three main scenes. Figure 5 shows the main scene in P-Soccer witth two buttons; `Play` and `Info`.

Figure 5. Main Interface of P-Soccer

Figure 6 shows the camera views for P-Soccer, which include the ball, goalpost, and spectator seats. The main character involved in this scene is the player who is taking the penalty kick. Finally, Figure 7 shows the third scene that includes the `Play` button.

Figure 6. Camera View of P-Soccer

Figure 7. Information Page

3.4. Implementation

All the assets involved in this application is organized in a file named 'KinectSoccer'. The mainframe for P-Soccer was developed using the Unity 3D. Figure 8 shows the model development in Unity 3D. Figure 9 shows the use of Adobe Flash CS6 to develop the buttons in P-Soccer.

Figure 8. P-Soccer Development in Unity 3D

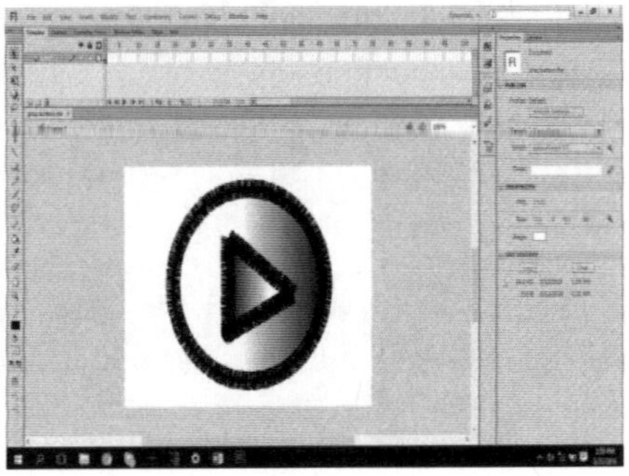

Figure 9. Interface Design in Adobe Flash CS6

4. Testing

Two types of application testing were applied to P-Soccer, which are alpha testing and beta testing. Alpha testing focused on button functionalities, properties setting, and response between the Kinect device and P-Soccer. Beta testing was carried out among the target users, who are varsity soccer players at Universiti Tun Hussein Onn Malaysia (UTHM), Malaysia. There were four groups involved based on player positions. Beta testing used the questionnaire as the medium to record the response. Two main components in the questionnaire are (1) gameplay and (2) functionality testing. Table 2 shows the responses for gameplay.

Table 2. Comparison of Application Technologies

	1	2	3	4	5
Application is easy to play				75%	25%
Background and properties suits the game theme				25%	75%
Characters are suitable and interesting				50%	50%
Music background has positive effects to the game			25%	50%	25%
Character animation is suitable and interesting				50%	50%
Multimedia elements are suitable with the game theme				25%	75%

According to the Likert scale, 1 being the highly disagree and 5 being highly agree. Using the same scale, Table 3 shows the results for functional testing. Overall, the target users responded well to the proposed P-Soccer games.

Table 3. Comparison of Application Technologies

	1	2	3	4	5
Tracking device is working properly					100%
Navigation buttons are working properly				25%	75%
Characters are easy to control				25%	75%
Screen quality is very good			25%	50%	50%

5. Conclusions and Future Works

This paper presented the development of a soccer games application using Kinect called the P-Soccer. P-Soccer focused on training for penalty kicks in soccer games by using the Kinect technology. P-Soccer introduces a new approach for interaction between users and the game characters. The interaction allows the game character to follow the body movement of the users. The main limitation in P-Soccer game design is the absence of goalkeeper. In real training, penalty kicks need the goalkeeper so the main character is able to learn the movement and how to respond accordingly. In addition to the goalkeeper, in the future, P-Soccer will be improved to allow the main character to change position when taking the penalty kick. This is to simulate in real training where players will determine their own distance to the goal post.

Acknowledgements

This project is sponsored by the Malaysian Ministry of Higher Education (MOHE) under the Research Acculturation Grant Scheme vot 1513.

IOP Conf. Series: Journal of Physics: Conf. Series **1020** (2018) 012005 doi:10.1088/1742-6596/1020/1/012005

References

[1] Berita Harian. Terokai Ensiklopedia Dunia. Available at https://udomworld.files.wordpress.com/2013/06/kenal-bola.pdf

[2] FIFA Official Website. Available at www.fifa.com-LawsOfPenaltyKick

[3] Gee J P 2003 What Video Games have to Teach us about Learning and Literacy. Computers in Entertainment (CIE) 20

[4] Kinect Hardware.: Available at http://electronics.howstuffworks.com/

[5] Darby J 2014 Beginning OpenGL Game Programming Cengage Learning

[6] Barton M and Loguidice B 2015 The History of Pong. Available at http://www.gamasutra.com/view/feature/3900/the_history_of_pong_avoid_missing_.php?print=1

[7] Kolbjornsen M 2012 A Comparison of Motion-Sensing Game Technologies for use in Physical Rehabilitation, Norwegian University Of Science and Technology. Master's thesis, Institutt for Datateknikk Gg Informasjonsvitenskap

[8] Kean S, Hall J and Perry P 2011 Meet the Kinect: An Introduction to Programming Natural User Interface Apress

IOP Conf. Series: Journal of Physics: Conf. Series **1020** (2018) 012006 doi:10.1088/1742-6596/1020/1/012006

Kinematics Analysis of Dominant and Non-Dominant Lower Limb during Knee Strike among MuayThai Beginners

Chamnan Chinnasee[1], Ali Md Nadzalan[2], Nur Ikhwan Mohamad[2], Abdul Hafiz Ahmad Sazili[2], Witthaya Hemapandha[1], Thariq Khan Azizuddin Khan[2], Luckhana Pimjan[3] & Kevin Tan[2]

[1]Faculty of Health and Sports Science, Thaksin University Papayom, Phatthalung, Thailand

[2]Faculty of Sports Science and Coaching, Sultan Idris Education University, Tanjung Malim, Perak, Malaysia

[3]School of Liberal Arts, Walailak University, Tasala, Nakhon Si Thammarat, Thailand

Corresponding author's emal: chamnan21@hotmail.com

Abstract. This study was conducted to determine and compare the kinematics of knee strike in MuayThai between dominant and non-dominant lower limb. Ten MuayThai beginners (mean age = 20 ± 1 years old) with less than one week experiences in MuayThai training were recruited and were asked to perform three trials of knee strikes for each leg (dominant and non-dominant). Joint angles and angular velocity of the movement were assessed for each trial. Repeated measure multivariate analyses of variances (MANOVA) were performed to compare the kinematics data between the dominant and non-dominant lower limb. Results showed no significant differences existed in all the joint kinematics examined between dominant and non-dominant lower limb. As the conclusion, MuayThai beginners demonstrated no differences of joint kinematics during knee strike between dominant and non-dominant lower limb.

1. Introduction

Thai Boxing or more popular known as MuayThai is a martial art and a method of self-defence. Started in the ancient times as a way to protect self from the enemies and even the animals [1], MuayThai now has become one of the most popular martial art sport in the world. A MuayThai match consisted of five rounds of fighting with three minutes per round and two minutes rest between rounds. Four offensive maneuvers are allowed to be used in MuayThai matches; punches, knee-strikes, elbow-strikes, and various kinds of kicks. All of these maneuvers can all be used to attack any part of the opponent's body.

Published studies in MuayThai has focusing on the injury analysis [2-5], physiological responses [6, 7], strength and conditioning [8], biomechanics [1, 9-11], performance analysis [12, 13] and psychology [14, 15]. A review by Mohamad [16] addressed the needs of more studies conducted on MuayThai for sports scientist and practitioners alike to develop systematic training and monitoring plan for MuayThai athletes.

In order to improve performance in a sport, it is important to analyse the movement especially the techniques that been allowed to be performed in a sport. One of the most important techniques

Content from this work may be used under the terms of the Creative Commons Attribution 3.0 licence. Any further distribution of this work must maintain attribution to the author(s) and the title of the work, journal citation and DOI.
Published under licence by IOP Publishing Ltd

performed in MuayThai is knee strike. The knee strike is a strike with the knee, either with the kneecap or the surrounding area. There are three types of knee strike; i) straight, ii) curved and ii) flying. Of all these, the straight knee strike is the most popular and always been performed in a match. In this study, the straight knee strike will be used interchangeably with knee strike. The knee strike involves the thrusting of the front knee into the head or body of an opponent. Knee strike always been performed when the performer is clinching the head or shoulder of the opponent. An athlete that able to perform the knee strike effectively will have the ability to defeat their opponent by landing accurate shots to opponents' body part such as the head, stomach, ribs, thighs, solar plexus and hips.

Instead of the importance of the knee strike, lack of studies has been conducted on the action. Studies on the biomechanics of MuayThai has been conducted on analysing the kinetic and kinematic parameters of the roundhouse kick [1], and the kinematics of the two Thai Boxing clinching positions [10]. Trial and Wu [10] in their study analyse the lower limb kinematic during knee strike in two clinching techniques; i) double collar tie and ii) double underhook. The findings of the study stressed the importance of hip and knee joint movements in both clinching positions.

It is important for the research to be conducted specific to the movement involved in a sport as the findings will demonstrate the mechanical demands during the techniques execution that will then add the information and knowledge for training program design. Besides that, it is also important for study to be conducted on both the dominant and non-dominant limb as this will demonstrate any asymmetry that existed [17-19]. This study was conducted among the Muay Thai beginners to look at if there are any differences in terms of kinematics data among the population so that correction can be done by the coach and individuals in order to enhance the functionality of both limb sites. Previous studies conducted on comparing dominant and non-dominant lower limb kinematics in martial art has been conducted on the roundhouse kick in taekwando [20-22], turning kick in taekwando [23], front kick in karate [24], and clinching in MuayThai [9]. Lack of studies has been conducted on the knee strike kinematics differences between both limb sites.

Thus, it is the objective of this study to determine and compare the joint kinematics of knee strike in MuayThai. The analysis will involve both dominant and non-dominant lower limb and comparison will be made between both sides of lower limb.

2. Methodology

2.1. Participants

10 male MuayThai beginners (mean age = 20 ± 1 years old) were recruited as study participants. Participants have less than 1 week experience in systematic MuayThai training. Participants had no medical problems and were screened prior to testing using PAR Q. Each participant had read and signed an informed consent for testing approved by the Universiti Pendidikan Sultan Idris and Thaksin University Ethics Committee.

2.2. Knee Strike

Participants were involved in a familiarization session to familiarize them with the techniques execution. All information about the data collection procedure was informed during the familiarization session. During the testing session, participants were first allowed of 10-15 minutes warm up session that consisted of stretching and kicking practiced conducted by a qualified national level MuayThai coach. After warm up, participants underwent their knee strike assessment. Participants were instructed to take two steps behind before execute the knee strike movement. During the starting (ready) position, participants were asked to face an instructor that were holding a kicking target, with knees that were slightly bend, and the knee that was going to kick at that trial need to be in front of the other (Figure 1a). The participant raised his hand up to the front as to simulate in the real conditions (hands holding the shoulder or head of the opponent) (Figure 1b). The participant next do a step forward (Figure 1c) before execute a knee strike to the kicking target (Figure 1d). Participants were reminded to perform the movement as hard and as fast as possible. Each participant was required to perform three trials for each leg that made up of total 6 trials for each participant.

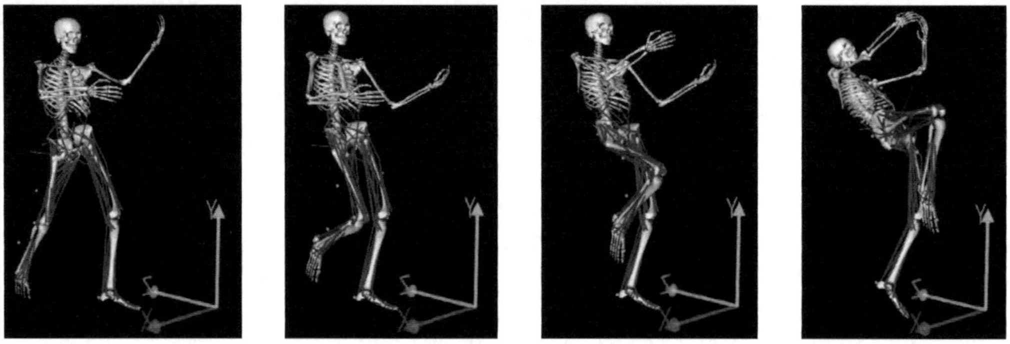

a b c d

Figure 1

2.3. Kinematics Assessment
Six infra-red cameras motion analysis system (Vicon T10s, Oxford Metrics, UK) was used to collect kinematics data, sampled at 100 Hz. Based on the Plug-in-Gate Marker Set, 29 reflective markers were attached to participant body at the second metatarsal, lateral malleolus, calcaneus, lateral shank, lateral femoral epicondyle, lateral thigh and anterior superior iliac spine at both sides of body. The kicking target used was two hand-held kicking pads (MUAY THAI MT-KPC 1, Thailand). The kicking target was held by an assistant with hands at the body level of the subject. Analysis of data was conducted using Vicon Workstation software. The kinematic model of the lower body consisted of the shank, thigh and pelvis of the kicking leg. The angles of the hip, knee and ankle were examined.

2.4. Statistical Analysis
Descriptive data and mean score were measured using descriptive analysis. Repeated measure analysis of variances (MANOVA) was used to compare the differences of kinematics data between both dominant and non-dominant limbs. α-level of $p < 0.05$ was accepted as the statistical significance. All statistical analyses were conducted using SPSS version 23 (IBM, New York, USA).

3. Results
Table 1 showed the physical characteristics of participants involved in this study.

Table 1: Physical characteristics of participants

Age (years)	Body mass (kg)	Height (cm)	Right Leg length (cm)	Left leg length (cm)
20 ± 1	65.33 ± 2.68	170 ± 5	85.65 ± 4.63	85.50 ± 4.48

Table 2 showed the mean angle of dominant and non-dominant knee during the contact. No significant main effect were found for the knee flexion/extension, $F(1,9) = 0.332$; $p > 0.05$, knee internal/external rotation, $F(1,9) = 0.870$; $p > 0.05$, knee abduction/adduction, $F(1,9) = 0.863$; $p > 0.05$.

Table 2: Knee joint angles at contact

Knee	Dominant (mean ± SD)	Non-dominant (mean ± SD)	p-value
Flexion/extension (°)	112.40 ± 10.35	108.94 ± 10.72	.578
Internal/external rotation (°)	23.48 ± 13.84	28.32 ± 17.62	.375
Abduction/adduction (°)	7.78 ± 15.81	2.18 ± 10.11	.377

Table 3 showed the mean angle of dominant and non-dominant hip during the contact. No significant main effect were found for the hip flexion/extension, $F(1,9) = 1.239$; $p > 0.05$, hip internal/external rotation, $F(1,9) = 0.824$; $p > 0.05$, hip abduction/adduction, $F(1,9) = 1.368$; $p > 0.05$.

Table 3: Hip joint angles at contact

Hip	Dominant (mean ± SD)	Non-dominant (mean ± SD)	p-value
Flexion/extension (°)	35.63 ± 18.39	42.68 ± 16.61	.295
Internal/external rotation (°)	-33.90 ± 16.34	-28.81 ± 11.52	.388
Abduction/adduction (°)	5.19 ± 20.98	13.09 ± 14.79	.272

Table 4 showed the mean angular velocity of knee and hip flexion for both dominant and non-dominant sites. No significant main effect were found for the knee flexion velocity, $F(1,9) = 0.019$; $p > 0.05$, hip flexion velocity, $F(1,9) = 0.714$; $p > 0.05$.

Table 4: Angular velocity at contact

	Dominant (mean ± SD)	Non-dominant (mean ± SD)	p-value
Knee flexion (°/s)	133.41 ± 62.22	143.46 ± 56.84	.893
Hip flexion (°/s)	258.89 ± 214.47	203.34 ± 55.42	.420

Table 5 showed the mean of knee forward and knee vertical linear velocity for both dominant and non-dominant sites. No significant main effect were found for the knee forward velocity, $F(1,9) = 0.211$; $p > 0.05$, knee vertical velocity, $F(1,9) = 0.508$; $p > 0.05$.

Table 5: Peak linear velocity

	Dominant (mean ± SD)	Non-dominant (mean ± SD)	p-value
Knee forward (m/s)	1.04 ± 0.90	1.15 ± 0.52	.657
Knee vertical (m/s)	4.15 ± 1.30	4.38 ± 1.43	.494

4. Discussions

Knee strike is one of major offensive techniques performed in MuayThai and it would be interesting to explore further on its techniques execution and effect. This study was conducted to determine and compare the joint angular displacement, velocity and acceleration of the hip, knee and ankle during knee strike for both dominant and non-dominant lower limb.

First of all, this study had found that no significant differences existed for all the kinematics comparisons made between dominant and non-dominant lower limb. Similar findings can be seen in other previous studies that had compared dominant and non-dominant lower limb kinematics in martial art [9, 20, 22-25].

No significant differences were also found for the angles of joints examined in this study. At first, it was thought that since knee strike techniques can be said as exclusively applied by MuayThai fighters only, with different kicking techniques compared to other types of kicks, the outcome may be different. But it seems this findings was consistent with several previous findings that had also found no significant differences of lower limb joint angles during turning kick in taekwando [23], front kick in karate [24], clinching in MuayThai [9] and roundhouse kick in taekwando [21, 22]. It is suggested that the similarities may be due to the nature of combat sports where regardless of dominant or non-dominant, both opponents need to be prepared and use both legs in defending and attacking movements.

Looking at the velocity data, no significant differences were found for all the angular and linear velocity examined in this study. This was partly in contrast to the findings of [9] that found joint angular velocity for the knee joint was faster for the dominant compared to the non-dominant leg during knee strike in two different clinching technique. Moreover, study by Hsieh, Huang [20] found that dominant leg is significantly better than non-dominant leg in the peak toe velocity, impact toe velocity and peak ankle linear velocity.

The explanation behind this contrast in findings could be caused by the differences of participants recruited in which this study involved beginners while Hsieh, Huang [20] and Trial [9] recruited highly experienced martial art practitioners. Different kinds of techniques execution could also contribute to the differences in which Hsieh et al. (2012) in their study conducted roundhouse kick in Taekwando which is a more complex movement compared to the knee strike in this current study.Knee strike could be said as a simple movement in which participant only need to raise the knee in front. Perhaps, the easiness of the technique has caused the participant to able to perform the knee strike with the non-dominant leg as efficient as the dominant leg.

When comparing the current data to the previous conducted by Trial [9], it can be seen that the velocity data obtained in this current study were far less than obtained in Trial [9]. The differences could be explained by the differences of participants recruited. Participants in this current study were beginners that just started MuayThai training for less than a week. The participants were believed to still have lack of motor abilities to perform the movement as efficient as the participants recruited by Trial [9] that were trained and have at least 5 years experienced in MuayThai training that are believed to already adapt and master the movement.

5. Conclusions

This study provides an important understanding on the kinematics of one of the most popular techniques performed in MuayThai, the knee strike in both the dominant and non-dominant lower limb. Ten MuayThai beginners were recruited and each athlete performed three trials for each sites. The findings of this study indicate no significant differences were found during knee strike between dominant and non-dominant lower limb. More biomechanics researches using advance technologies are warranted to be done on MuayThai technique as the speed of the martial art movements were very high make them virtually impossible to be analysed with the naked eye, and many traditional assumptions about optimal techniques may be incorrect.

References

[1] Sidthilaw, S., *Kinetic and kinematic analysis of Thai boxing roundhouse kicks.* 1996.

[2] Gartland, S., M. Malik, and M. Lovell, *Injury and injury rates in Muay Thai kick boxing.* British Journal of Sports Medicine, 2001. 35(5): p. 308-313.

[3] Gartland, S., M.H. Malik, and M. Lovell, *A prospective study of injuries sustained during competitive Muay Thai kickboxing.* Clinical Journal of Sport Medicine, 2005. 15(1): p. 34-36.

[4] Zazryn, T.R., C.F. Finch, and P. McCrory, *A 16 year study of injuries to professional kickboxers in the state of Victoria, Australia.* British Journal of Sports Medicine, 2003. 37(5): p. 448-451.

[5] Vaseenon, T., et al., *Foot and ankle problems in Muay Thai kickboxers.* Journal of the Medical Association of Thailand= Chotmaihet thangphaet, 2015. 98(1): p. 65-70.

[6] Crisafulli, A., et al., *Physiological responses and energy cost during a simulation of a Muay Thai boxing match.* Applied Physiology, Nutrition, and Metabolism, 2009. 34(2): p. 143-150.

[7] Cappai, I., et al., *Physiological responses and match analysis of Muay Thai fighting.* International Journal of Performance Analysis in Sport, 2012. 12(3): p. 507-516.
[8] Turner, A.N., *Strength and conditioning for Muay Thai athletes.* Strength & Conditioning Journal, 2009. 31(6): p. 78-92.
[9] Trial, W., *Kinematics Comparison between dominant and non-dominant lower limbs in Thai Boxing.* 2013.
[10] Trial, W. and T. Wu, *A kinematic analysis of the Thai boxing clinch.* Advances in Biomechanics and Applications, 2014. 1(1): p. 57-66.
[11] Chinnasee, C., et al., *Lower limb kinematics analysis during roundhouse kick among novices in Muay Thai.* Journal of Fundamental and Applied Science, 2017. 9(6S): p. 1002-1010.
[12] Ouergui, I., et al., *Time-motion analysis of elite male kickboxing competition.* The Journal of Strength & Conditioning Research, 2014. 28(12): p. 3537-3543.
[13] Silva, J.J.R., et al., *Time-motion analysis in Muay-Thai and kick-boxing amateur matches.* 2011.
[14] Myers, T.D. and N.J. Balmer, *The impact of crowd noise on officiating in Muay Thai: achieving external validity in an experimental setting.* Frontiers in Psychology, 2012. 3: p. 346.
[15] Ong, T.F. and W.I. bin Wan Ruzmin. *Participation motivation in Muay Thai among Malaysians*. in *Proceedings of the 2nd International Colloquium on Sports Science, Exercise, Engineering and Technology 2015 (ICoSSEET 2015)*. 2016. Springer.
[16] Mohamad, N.I., et al., *Sport science based research on the sports of Muay Thai: A systematic review.* Walailak Journal of Science and Technology (WJST), 2016. 14(8).
[17] Nadzalan, A.M., et al., *Muscle activation analysis of step and jump forward lunge among badminton players.* Journal of Engineering and Science Research, 2017. 1(2): p. 60-65.
[18] Nadzalan, A.M., et al., *Fascicle behaviour analysis during forward lunge exercise: the comparisons between training loads.* Journal of Fundamental and Applied Science, 2017. 9(6S): p. 1090-1101.
[19] Nadzalan, A.M., et al., *Kinetics analysis of step and jump forward lunge among badminton players.* Journal of Fundamental and Applied Science, 2017. 9(6S): p. 1011-1023.
[20] Hsieh, A., C.-F. Huang, and C.C. Huang. *The biomechanical analysis of roundhouse kick in taekwondo*. in *ISBS-Conference Proceedings Archive*. 2012.
[21] Falcó, C., et al. *Kinetic and kinematic analysis of the dominant and non-dominant kicking leg in the taekwondo roundhouse kick*. in *ISBS-Conference Proceedings Archive*. 2009.
[22] Tang, W., J. Chang, and Y. Nien, *The kinematics characteristics of preferred and non-preferred roundhouse kick in elite Taekwondo athletes.* Journal of Biomechanics, 2007. 40: p. S780.
[23] Harun, H. and S.J. Xiong, *The symmetry in kinematics between the dominant and non-dominant legs in taekwondo turning kick.* Unpublished. Degree thesis, UniversitiTeknologi Malaysia, Skudai, Malaysia, 2010.
[24] Andrzejewski, X. and L. Elbaum. *Biomechanical analvsis of the front kick with the dominant and non-dominant limb in the shito-rvu style of karate*. in *ISBS-Conference Proceedings Archive*. 2008.
[25] Falco, C., et al., *Influence of the distance in a roundhouse kick's execution time and impact force in Taekwondo.* Journal of Biomechanics, 2009. 42(3): p. 242-248.

IOP Conf. Series: Journal of Physics: Conf. Series **1020** (2018) 012007 doi:10.1088/1742-6596/1020/1/012007

Repeated Sprint Ability with Inclusion of Changing Direction among Veteran Soccer Players

Omar Md Salleh[1], Ali Md Nadzalan[1], Nur Ikhwan Mohamad[1], Azali Rahmat[1], Mirza Azny Mustafa[1], & Kevin Tan[1]

[1]Faculty of Sports Science and Coaching, Sultan Idris Education University, Tanjung Malim, Perak, Malaysia

Corresponding author's emal: omar@fsskj.upsi.edu.my

Abstract. This study was conducted to determine the repeated sprint ability (RSA) with the inclusion of changing direction among veteran soccer players. Twelve main players from a university veteran soccer team were recruited and were required to perform the RSA test in two conditions; i) without ball and ii) with ball. Descriptive statistics and Wilcoxon signed rank test was conducted to determine the mean score and the differences of sprint time, percentage of decrement score of sprinting (Sdec) and fatigue index (FI) between the two conditions. Both conditions demonstrated significantly drop in speed in the fifth sprint. Results showed sprint time, Sdec and FI were found to be significantly different between both conditions. Findings of this study demonstrated the important for specific training to sports for performance enhancement.

1. Introduction

Studies on improving physical abilities among soccer players has been expanding and keep been conducted over the years [1-3]. Several previous studies showed an athlete from a team sport is required to sprint for 1-10% of the total distance covered during the actual game that is almost 1-3% playing time [4-7]. The findings of these studies demonstrated the importance of team sports' athlete to have good ability to sprint repeatedly in order to enhance ball possession for their team.

In accessing the repeated sprint ability (RSA), various protocols of test existed and it is a matter of which test is more specific to be conducted among athletes. Repeated sprint ability (RSA) is defined as the capability to reproduce maximum speed over a series of sprints [8]. Study by Rampinini, Bishop [9] has shown the average time recorded during an RSA test predicted the distance of high-intensity running (>19.8km/h), and the total sprint distance during a professional soccer match. Thus, it can be seen that RSA development can positively affect performance of a team.

Several methods have been examined and have found to improve RSA [10-15]. However, before starting a test, it is important to assess the performance of the athletes so that the initial level of the players can be obtained besides can know the suitability of the tests conducted.

In conducting RSA test or training, one of the things to be considered is the fatigue developed during the exercise as this will reveal the athlete's ability to withstand fatigue. Fatigue in RSA refer to the decrement of the maximal power output or speed (i.e. during cycling or running), although the individuals can sustained the task [14]. Fatigue during RSA training has been shown can rapidly develop as early as just after the first sprint Mendez-Villanueva, Hamer [16].

Content from this work may be used under the terms of the Creative Commons Attribution 3.0 licence. Any further distribution of this work must maintain attribution to the author(s) and the title of the work, journal citation and DOI.
Published under licence by IOP Publishing Ltd

It is important to reduce the fatigue development during games as the ability to produce maximum sprints may increase the chance to win ball possession that may positively affect the final outcome of a game. However, the inability to produce maximum speed during sprinting will increase the risk of losing ball possession and even concede goals. For example, study by Paton, Hopkins [17] have demonstrated that a ~0.8% reduction in sprint speed would increase the likelihood of a player losing possession of the ball against an opponent, when both players sprint for the ball.

Despite increasing number of scientific studies on RSA [5, 9, 18-21] lack of data existed on the RSA among veteran athletes. It is unknown on how the performance of veteran athletes in RSA causing the training program for them always been in general form. It was the aim of this study to determine the repeated sprint ability (RSA) with the inclusion of changing direction among veteran soccer players. The RSA was analysed through the sprint time, percentage decrement score of sprinting (Sdec) and fatigue index (FI).

2. Methodology

2.1. Participants

Twelve university veteran football representatives (mean age = 44.69 ± 2.03 years old) were recruited for this study. Participants were currently active training three times per week and were in their preparation phase for inter-varsity competition. PAR Q was used to screened participants prior to testing. Each participant has read and signed an informed consent for research participation approved by the Research Management and Innovation Centre, Universiti Pendidikan Sultan Idris (Code: RAGS/2013/UPSI/SG/01/2). All participants were informed on the study aims and procedures before the data collection.

2.2. RSA Procedures

Figure 1. RSA test set-up developed for this study [image taken from Salleh, Nadzalan [22]]

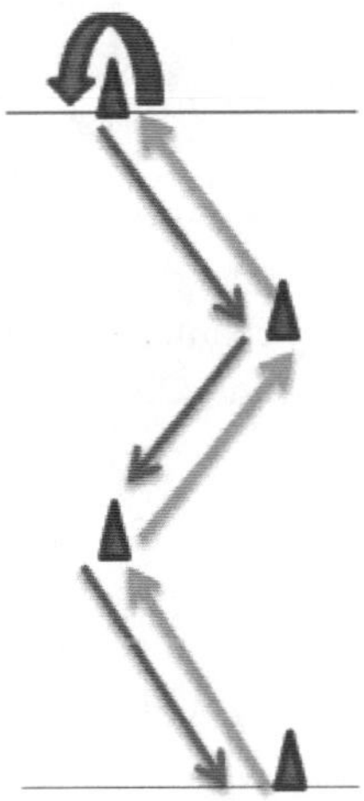

Figure 1 showed the RSA test set-up developed for this study. The RSA test involved 5 repetitions of maximal effort 40 m run with 60 seconds active rest in between. Participants were needed to run 6 m (60° straight to the left) before make the first turn (60° straight to the right for 7 m) and then make the second turn (60° straight to the left for 7 m) and continue to the end line before return back. During the return back, participants made 180° turn and also need to run through the two turns to go back to the starting point. Markers were placed at all the points that participants need to make turns.

One minute recovery consisted of 40 seconds walk around and twenty seconds before the start of next sprint, participants need to get ready at the starting line and await the start signal from an instructor. Participants were instructed to stand passively with non-dominant leg were used as a leading foot at the starting line. 'Ready, Set, Go!' was used as the start signal involve the instruction of the instructor.

Two RSA test were needed to be performed by all participants; one without ball and one with ball. For with ball RSA test, participant need to sprint while dribbling the ball until reached the 20 m point where they were needed to kick the ball to a small goal that had been set up 7 m straight away from the 20 m point. Participants just need to sprint back to the starting line without ball.

A timing gate (Microgate, Bolzano, Italy) was used to measure the sprint time the timing gate was placed at the starting line, 1 m above the ground. The investigators used a hand-held Q&Q Quartz stopwatch (Citizen Watch Co., Ltd., Tokyo, Japan) to monitor recovery time. The participants commenced each sprint, starting from a standing position 0.5 m behind the sensor. Strong verbal encouragement was provided to each participant during all sprints. Three scores were calculated for the analysis; i) sprint time, ii) FI, and iii) Sdec.

2.3. Formula

Both the formula of FI (eq. 1) and Sdec (eq. 2) were adapted from Girard, Mendez-Villanueva [14]. Sdec was measured to quantify fatigue by comparing actual performance to an imagined 'ideal performance' while FI was measured to indicate the drop-off in performance from the best to worst sprint performance [14].

Equation1

$$FI = [(S_{best} - S_{worst}) / S_{best}] \times 100$$

Equation 2

$$Sdec\ (\%) = [(S_1 + S_2 + S_3 + S_4 + S_5) / (S_{best} \times \text{number of sprints}) - 1] \times 100$$

2.4. Statistical Analysis

Normality of data was examined using Shapiro-Wilk test. The mean and standard deviation of participants' physical characteristics, sprint time, Sdec and FI were analysed using descriptive statistics while Wilcoxon signed rank test was conducted to assess the different of sprint time, FI and Sdec of the RSA test performed with ball and without ball.

3. Results

Table 1 showed the physical characteristics of participants.

Table 1. Physical characteristics of participants

	Mean ± SD
Age (years)	44.69 ± 2.03
Body mass (kg)	70.18 ± 5.69
Height (cm)	170.25 ± 4.38

Table 2 showed the time for each sprint in RSA without ball conditions. Sprint 1 and Sprint 2 were shown to be significantly faster compared to Sprint 5 (Sprint 1, $p = 0.01$; Sprint 2, $p = 0.021$).

Table 2. Sprint time without ball

Sprint	RSA time without ball (s) Mean ± SD
Sprint 1	8.77 ± 0.83
Sprint 2	8.84 ± 0.99
Sprint 3	8.88 ± 0.98
Sprint 4	8.89 ± 0.84
Sprint 5	9.15 ± 1.10

Table 3 showed the time for each sprint in RSA with ball conditions. Sprint 1 and Sprint 2 were shown to be significantly faster compared to Sprint 5 (Sprint 1, $p = 0.04$; Sprint 2, $p = 0.04$).

Table 3. Sprint time with ball

Sprint	RSA time with ball (s) Mean ± SD
Sprint 1	9.96 ± 0.96
Sprint 2	9.99 ± 1.09
Sprint 3	10.07 ± 1.01
Sprint 4	10.14 ± 0.99
Sprint 5	10.14 ± 1.21

Table 4 showed the comparison of average sprint time, FI and Sdec between RSA without ball and with ball. Results showed the sprint time without ball was significantly faster compared to sprint time with ball, $p < 0.01$. Besides that, FI and Sdec were found to be significantly higher in with ball condition, $p < 0.01$.

Table 4. Comparison of average sprint time, FI and Sdec between RSA without ball and with ball

	Time (s) Mean ± SD	FI Mean ± SD	Sdec Mean ± SD
RSA without ball	8.90 ± 0.92	7.15 ± 4.06	2.87 ± 1.19
RSA with ball	10.06 ± 1.03	5.15 ± 2.90	2.43 ± 1.46
% difference	13%	27.97%	15.33%
sig	0.002	0.002	0.002

i) Time = average sprint time, ii) FI = fatigue index, iii) Sdec = percentage decrement score of sprinting, iv) % difference = percentage difference of score between without ball and with ball, v) sig = p-value.

4. Discussions

Repeated sprint ability with the inclusion of changing direction among veteran soccer players were examined as the purpose of this study. Participants performed the RSA test in two conditions; one without ball and one with ball.

In without ball condition, Sprint 1 and Sprint 2 were shown to be significantly faster compared to Sprint 5 (Sprint 1, $p = 0.01$; Sprint 2, $p = 0.021$). This showed that participant didn't able to maintain their speed in the fifth sprint thus causing it to be significantly slower compared to the earlier sprints. Table 3 showed the time for each sprint in RSA with ball conditions. Again, Sprint 1 and Sprint 2 were shown to be significantly faster compared to Sprint 5 (Sprint 1, $p = 0.04$; Sprint 2, $p = 0.04$). As in the without ball condition, this also showed that participants sprint performance drop significantly in the last sprint.

Table 4 showed the comparison of average sprint time, FI and Sdec between RSA without ball and with ball. Researchers have tended to use two terms as a way to quantify the ability to resist fatigue during RSE, which are; i) fatigue index (FI) and/or ii) the percentage decrement score (Sdec) [14]. The FI has generally been calculated as the drop-off in performance from the best to worst sprint performance during an RSE whereas the Sdec attempts to quantify fatigue by comparing actual performance to an imagined 'ideal performance' (i.e. where the best effort would be replicated in each sprint) [14].

Results in this study showed the sprint time without ball was significantly faster compared to sprint time with ball, $p < 0.001$. Besides that, FI and Sdec were found to be significantly higher in with ball condition, $p < 0.01$. As in Salleh, Nadzalan [22], this finding also found that without ball condition produce greater sprint ability. This demonstrated no matter whom the population is, the sprint performance drop when in with ball condition. Findings also found FI to be higher, demonstrated that participants easier to become fatigue when in with ball condition.

Overall, this study demonstrated the lack of sprint ability and less fatigue-resistance when running with ball. This condition is a thing that should be taken care, as it is the ball control that is important in a soccer match. The inability to move better when in ball position inflicts the cases of ball possession that might occur in real game.

It is stressed that every team and players to enhanced their capability to sprint with ball efficiently. Training should be performed to enhance this aspect. Many types of training can be performed to improve RSA and among of them are the intermittent-sprint, repeated-sprint, speed, agility and quickness (SAQ) and coordination training [10-14].

5. Conclusions

To conclude, it can be seen from this study that the ability to sprint repeatedly were depending on the level of condition. It is important to be noticed that the RSA procedure used in this study has taken into account the specific requirement of soccer players in the real game. Future studies were suggested to be conducted on the high level soccer players (i.e. national level) and the methods to reduce FI and Sdec such as training method and dietary intake.

References

[1] Eniseler, N., et al., *High-intensity small-sided games versus repeated sprint training in junior soccer players.* Journal of Human Kinetics, 2017. 60(1): p. 101-111.

[2] Franco-Marquez, F., et al., *Effects of combined resistance training and plyometrics on physical performance in young soccer players.* International Journal of Sports Medicine, 2015. 94(11): p. 906-914.

[3] Md Noor, A., et al., *The effectiveness of "Small, Medium and Big Games" training on cardiovascular endurance among university football llayers.* International Journal of Academic Research in Progressive Education and Development, 2017. 6(4): p. 157-162.

[4] Buchheit, M., et al., *Repeated-sprint sequences during youth soccer matches.* International Journal of Sports Medicine, 2010. 31(10): p. 709-716.

[5] Spencer, M., et al., *Physiological and metabolic responses of repeated-sprint activities.* Sports Medicine, 2005. 35(12): p. 1025-1044.

[6] Spencer, M., D. Bishop, and S. Lawrence, *Longitudinal assessment of the effects of field-hockey training on repeated sprint ability.* Journal of Science and Medicine in Sport, 2004. 7(3): p. 323-334.

[7] Stølen, T., et al., *Physiology of soccer.* Sports Medicine, 2005. 35(6): p. 501-536.

[8] Bishop, D., O. Girard, and A. Mendez-Villanueva, *Repeated-sprint ability—Part II.* Sports Medicine, 2011. 41(9): p. 741-756.

[9] Rampinini, E., et al., *Validity of simple field tests as indicators of match-related physical performance in top-level professional soccer players.* International Journal of Sports Medicine, 2007. 28(03): p. 228-235.

[10] Milanović, Z., et al., *Effects of a 12 week saq training programme on agility with and without the ball among young soccer players.* Journal of Sports Science & Medicine, 2013. 12(1): p. 97.

[11] Venturelli, M., D. Bishop, and L. Pettene, *Sprint training in preadolescent soccer players.* International Journal of Sports physiology and Performance, 2008. 3(4): p. 558-562.

[12] Balsom, P., et al., *Maximal-intensity intermittent exercise: effect of recovery duration.* International Journal of Sports Medicine, 1992. 13(07): p. 528-533.

[13] Duffield, R., M. King, and M. Skein, *Recovery of voluntary and evoked muscle performance following intermittent-sprint exercise in the heat.* International Journal of Sports Physiology and Performance, 2009. 4(2): p. 254.

[14] Girard, O., A. Mendez-Villanueva, and D. Bishop, *Repeated-sprint ability—Part I.* Sports Medicine, 2011. 41(8): p. 673-694.

[15] Bishop, D. and B. Claudius, *Effects of induced metabolic alkalosis on prolonged intermittent-sprint performance.* Medicine and Science in Sports and Exercise, 2005. 37(5): p. 759-767.

[16] Mendez-Villanueva, A., P. Hamer, and D. Bishop, *Fatigue in repeated-sprint exercise is related to muscle power factors and reduced neuromuscular activity.* European Journal of Applied Physiology, 2008. 103(4): p. 411-419.

[17] Paton, C.D., W.G. Hopkins, and L. Vollebregt, *Little effect of caffeine ingestion on repeated sprints in team-sport athletes.* Medicine and Science in Sports and Exercise, 2001. 33(5): p. 822-825.

[18] Bravo, D.F., et al., *Sprint vs. interval training in football.* International Journal of Sports Medicine, 2008. 29(8): p. 668-674.

[19] Impellizzeri, F., et al., *Validity of a repeated-sprint test for football.* International Journal of Sports Medicine, 2008. 29(11): p. 899.

[20] Perrey, S., et al., *Neural and muscular adjustments following repeated running sprints.* European Journal of Applied Physiology, 2010. 109(6): p. 1027-1036.

[21] Spencer, M., et al., *Time–motion analysis of elite field hockey, with special reference to repeated-sprint activity.* Journal of Sports Sciences, 2004. 22(9): p. 843-850.

[22] Salleh, O.M., et al., *Repeated sprint ability depending on the level of condition among university soccer players.* International Journal of Academic Research in Business and Social Sciences, 2017. 7(6): p. 506-514.

Development of Android Application for Measuring Cardiovascular Endurance Fitness for Military Cadet Officers

Mohar Kassim[1], Ahmad Mujahid Ahmad Zaidi[2] and Rahmat Sholihin Mokhtar[3]

[1] Director,Centre for Coaching Science, UPNM
[2] Dean of Academic Management, UPNM
[3]Centre for Coaching Science, UPNM

corresponding author's e-mail address: mohar@upnm.edu.my

ABSTRACT
Mobile software application has become a part of today's lifestyle. This mobile app is designed to help society to be physically active. The application is named UPNM Cardio Fitness, and is developed on the Android platform. The original purpose of the application is to measure and analyse the level of cardiovascular fitness of 18 years old male Military cadet Officers through a 2.4 km run test. The application is based on a data base using Google Fusion Table that stores and analyses the data received. The application consists of two parts: information of the individual and their respective fitness norms that can be accessed either automatically or manually. The classification of the norms is obtained from the fitness norms of 120 male cadets aged 18 years old. The norms are grouped into five categories which are: Excellent, Very Good, Good, Moderate and Poor. The software consists of 5 hyperlinks which are the main page, individual information, test result, file and record. The application is created using MIT App Inventor Software and Windows 7. The creation of the application has enabled researchers particularly in the Science Training programme in UPNM to carry out tests as well as to identify the level of fitness of their trainees immediately, accurately, and systematically.

***Keywords**: 2.4km run test, android, cadet student, cardiovascular endurance UPNM, google fusion, MIT App Inventor, mobile application, physical fitness norm, Windows 7*

1. Introduction

Physical fitness is a component that is used all the time to carry out an activity. Physical fitness is defined as one's condition that makes him able to undertake a daily task without feeling tired when maximum energy has been used, and there is still an extra energy for use during leisure time and in an emergency situation. This definition is supported by Penny & Clarke (2005) who state that physical fitness is the ability to carry out daily activities efficiently and energetically, without feeling extreme fatigue, and still have enough energy for recreational activities. This means that fitness is extremely important for every individual, especially in this age of rapid development that requires an individual to be constantly active and be at the optimum level of health. This scenario is in line with the advent of modern science and technology, which has become increasingly advanced in this millennium.

Content from this work may be used under the terms of the Creative Commons Attribution 3.0 licence. Any further distribution of this work must maintain attribution to the author(s) and the title of the work, journal citation and DOI.
Published under licence by IOP Publishing Ltd

IOP Conf. Series: Journal of Physics: Conf. Series **1020** (2018) 012008 doi:10.1088/1742-6596/1020/1/012008

Information Communication and Technology (ICT) revolves all aspects of technology as the medium of information dissemination such as radios, televisions, smart phones, computers and internet. Rapid advancement in ICT creates a lot of impacts in almost every aspects of life. It is used in broadcasting news, sharing educational knowledge including updates on health issues. The development in the field of technology has changed ways human being learn and work [1]. This is due to the globalisation of technology that leads it to be perpetually and continually developed throughout time. Combining the creativity and innovation of the researcher, the mobile app is introduced as something new that has never been created before. This is because technology has been used as a bridge between individual experience and existing knowledge, with new knowledge that is taught and learned through the delivery of mobile application used [2].

This study states that there is a requirement for the existence of a certain norm that can measure the cardiovascular endurance among each individual cadet. Until now, there is no means to measure the cardiovascular fitness to evaluate the physical health status of UPNM cadet students. For now, the administration of UPNM uses the physical fitness tests taken from AAHPERD (1996). The norm is based on conditions that do not fit the local people and most certainly has protocol and circumstances that are different from ours. Therefore, to best evaluate the fitness performance level of cadet students, there is a need for a norm that is valid and reliable in accordance to the protocol and medium suited to the setting of our country. Cardiovascular fitness norm refers to the order or numerical range that shows the physical fitness level of an individual based on a certain guideline. This norm is used to determine a specific fitness level after the completion of a fitness test.

Based on the current development in technology, as predicted, smartphones have become an indispensable communication tool for most people, especially youngsters [6]. The prevalence of using mobile technologies especially smartphones has shown a rapid speed in the ubiquities. This is supported by the convenience of the internet infrastructure for high speed internet access and WIFI facilities installed in most buildings. People are compelled to become digital citizens by having smartphones. Apart from the infrastructures provided, mobile application tools have also played a role in the increase use of smartphones. Smartphones are not purely designed to meet basic needs such as socialising using social applications, but it is designed to be more user friendly whereby anyone can use it despite not being IT savvy [7].

1.1. Research Objectives

This study was carried out to achieve the following objectives:

1.1.1 To develop a cardiovascular fitness norm for UPNM Military Cadet Officers according to their age and gender based on test scores of the mentioned norm.

1.1.2 To evaluate the physical fitness level of UPNM Military Cadet Officers based on a number of criteria fixed.

1.1.3 To build portable application software based on the cardiovascular fitness level of UPNM for Military Cadet Officers.

In this study, detailed planning was required in order for the above objectives to be achieved. Besides, the concepts of this study, which are portable application software and cardiovascular fitness norm, have to be made clear so that no complications can interfere with the achievement of the study.

1.2. Significance of the Study

The translation of data into an application that is being done in the research is something that has not been done before. The app is hoped to aid lecturers and physical trainers in managing as well as administrating a physical fitness test. The new system will inevitably improve the quality of the pre-existing management system. In addition, the creation of the app is crucial in managing the test results as there is no systematic way to save the information thus far. The app which is a computerized system helps other researchers to record, analyse and save the information of the students' physical fitness data to be stored online. The cadets can also find out quickly of their performance or the norms achieved from the test conducted.

1.3. Research Instrument

Cardiovascular Endurance Fitness Norms:
In order for the researcher to assess the level of physical fitness of a person in a study, there should be a test that is consistent with the subject matter of the study. The physical fitness test carried out in a study is considered as an instrument. Instruments in a research for a physical fitness test must have high validity and reliability in order for the result to be precise and indisputable [12], [13]. In this study, the chosen instrument is a 2.4 km run test developed by Kenneth H. Cooper in 1968 for the US military [14]. The 2.4 km run test has high validity, r=0.92 [15] and r=0.86 [16]. After the 2.4km run test is carried out, the results are then analysed. The cardiovascular endurance fitness norms produced will then be the guideline data to build the mobile software application that uses Android as the platform. The norms are to be accessed via smartphones.

Smartphone and Mobile Application:
The smartphone is a combination of mobile phone and personal digital assistant (PDA). The mobile phone uses an operating system such as Symbian, Windows Mobile, Android, iOS, Palm, and numerous mobile softwares. It comes with internet access and is able to support multimedia applications [17]. The 2 in 1 concept used in smartphones is useful to many users especially those who are frequently engaged in outdoor activities such as attending seminars or meetings, taking data of patients, retrieving serial number in hardware store, organising daily routine, finding direction or for navigation purposes and many more. For the PDA to function, smartphones are equipped with a microprocessor and memory chips such as RAM, ROM or flash card [18].

The operating system of a smartphone represents the producers. The two profound operating systems are Google's Android and Apple's iOS. Both of the operating systems have their own unique characteristics that represent the producer or company. Statistics has shown that smartphones using Android as the platform are more prevalent than others with a market share of 82% worldwide [19]. The chart below shows the differences between Android market and other operating systems.

2. Methodology

CONSTRUCTION OF CARDIOVASCULAR FITNESS NORMS
The research is done through experimental observation. It is designed to use the pre-existing instrument developed by Cooper Institute. Descriptive statistics analysis through SPSS version 20 was used to calculate and interpret the data. The findings have enabled the researcher to describe the frequency, percentage, mean and standard deviation of the existing demographic variable such as sex, race, height and weight. The samples of the study are 18 year old UPNM Foundation Programme male cadets for admission in the 2013/ 2014 session. A total of 120 male cadets carried out a fitness run test of 2.4 km representing 30 per cent of the total male cadets in the foundation programme [22].

2.1. Application Production

The cardiovascular endurance norms produced are then used as the main data for the production of the mobile application. The method of production of the mobile application is using the System Development Life Cycle (SDLC) model. It is a form of a database system development cycle and operation of the software. According to [23], the SDLC model consists of 6 phases, and they are: preliminary investigation, system analysis, system design, system development, system deployment and maintenance phase. These phases are cycles for the working of a process that will identify the strength or weakness of every phase. The first phase which is the preliminary investigation phase is the phase for producers of the app to identify the problems and needs for information systems. In the first phase of this research, the producer has identified and recognised that there is still no application system created specifically in recording, saving and managing the cardiovascular endurance physical fitness test information of cadets.

The second phase is the system analysis whereby the system analyses new requirements which a specified system needs after the current system is studied in depth. This phase is the most crucial phase because developers can develop the application based only on its needs to carry out the process easily as planned. The hardware required in this study is a computer and a mouse whereas the software required is online MOT Apps Inventor and an internet connection. MIT App Inventor is used because it is done virtually in the cloud by the website http://ai2.appinventor.mit.edu/ and does not require an installation of the app in the computer system. Problems and needs of the end users will be analysed in more detail so that the goals can be achieved.

The third phase is the application design phase. There are three main tasks in this phase, and they are: providing alternative application, choosing the best application design, and writing an application design report. The application producer will create more than one design to fulfil the need for information. Required specs from the previous phase will be studied and the application design will be prepared in this phase. The desired feature and operation will be described in detail including screen display, functions, hardware and software.

The fourth phase is the application development phase. In this phase, the application is developed using the software and hardware. The application is then put into work to test its functionality. A test typically takes about 2 weeks especially if it is complex. All feedbacks and responses are taken to ensure all procedures have worked out as planned.

The application deployment phase is the fifth phase which is the installation of the new system. End users will be taught and guided in using the application. Another name for this phase is the application conversion because there are conversion processes or changes from the older system to the new one. End users will be trained to use the new application accordingly.

The last phase is the maintenance phase whereby maintenance begins with updating the application. Evaluation is then made to see whether the application is productive and dynamic.

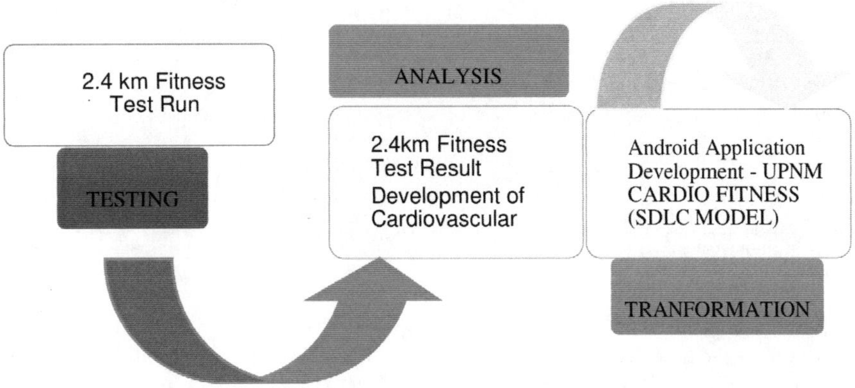

Figure 1: Research Framework

3. Results

All respondents were randomly selected for this study. Descriptive statistics showed mean scores and standard deviations for all respondents involved in the 2.4 km run test (M=10.86, SD=0.91). Table 1 shows the test scores of the students in the 2.4 km run test. The minimum and maximum time recorded is 9.17 and 13.13 mins while the median and the skewness showed 11.13 and -0.036. The statistics/ data produce a bell-shaped curve to show normal distribution.

Table 1: Descriptive Statistics for UPNM Foundation Programme Male Cadets

18 Years Old Male Military Cadets Officers	
Minimum	9.17
Maximum	13.13
Mean	10.8646
Median	11.1350
Std. Deviation	.91960
Skewness	-.036
18 year old Male	
N	120

Table 2 shows the test result of the norms from the 2.4 km run test conducted on the male cadets. There are 5 predefined classifications in the production of a particular grade of fitness norms, and they are: Excellent (5), Very Good (4), Moderate (3), Satisfactory (2), Poor (1). According to [12], grading a test enables a set of people to be classified according to their abilities and achievement. Every level states the scores achieved starting from the highest to the lowest which is 5 to 1. Norms are also used by the National Physical Fitness Award Singapore (NAPFA) [24.] From the table, the highest achievement which is Excellent with the score of 5 starts from 9:40 and below, the next level is Very Good with the scores of 4 is from 9:41 to 10:40, the Good level with the score of 3 is from 10:41 to 11:31, and the satisfactory level with the score of 2 is from 11:31 to 12:24 and lastly, the Poor level with the score of 1 is from 12:25 and above.

Table 2: Cardiovascular Endurance Physical Fitness Norms for 18 Years old UPNM Male Military Cadet Officers

Category	Score	Duration (mm:ss)
Excellent	5	< 9:40
Very Good	4	9:41 - 10:40
Good	3	10:41 - 11:30
Satisfactory	2	11:31 - 12:24
Poor	1	12:25 >

Note. (mm:ss)=(minutes:seconds)

4. Application Development

The application is named 'UPNM Cardio Fitness'. The app is produced to measure the level of cardiovascular endurance of 18 years old UPNM male Military Cadet officers. The application is produced using a software programme developed by Google which is then taken over by MIT. This software can be accessed through gmail account via website which is http://ai2.appinventor.mit.edu/. The researcher created a gmail account specifically for this study which is norm.upnm@gmail.com. The account is officially dedicated to the construction of the mobile app. Figure 2 shows the front page of the software used to develope the 'UPNM Cardio Fitness' mobile app.

The researcher was required to create a specific file using .aia format in the process of building the app. By using the format, the application is developed by following the guidelines in the app's construction framework. Tutorials are provided in building the app by MIT. They have prepared the process and procedure for those who intend to create their own application. The tutorials have facilitated the researcher in building the app. Interestingly; MIT App Inventors 2 does not require an Android emulator to run as the application player. The software uses an emulator developed by MIT AI2 Companion itself that can be downloaded straight to the smartphone. Hence, the researcher could run and test the application at the same time.

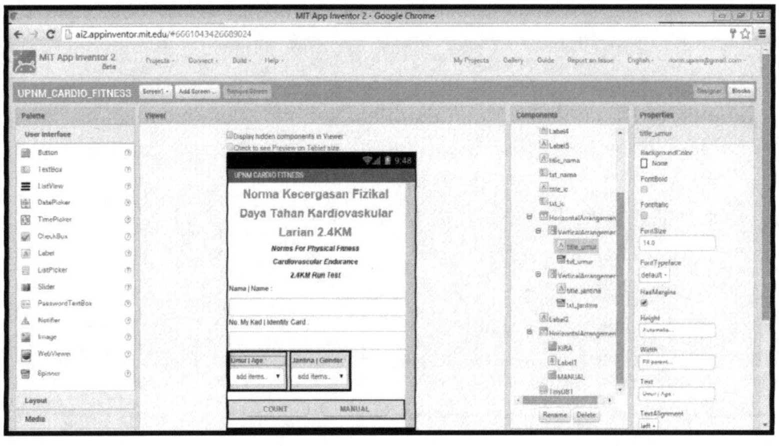

Figure 2: Interface features of MIT Apps Inventor through the UPNM_CARDIO_FITNESS.aia file

IOP Conf. Series: Journal of Physics: Conf. Series **1020** (2018) 012008 doi:10.1088/1742-6596/1020/1/012008

The researcher then changed the format from .aia to .apk to test the application. The application is downloaded from the registered account into the smartphone. The smartphone will open the UPNM CARDIO Fitness application. Figure 3 below shows the UPNM Cardio Fitness app that has been installed into a smartphone.

Figure 3: UPNM Cardio Fitness app installed into a smartphone

After the installation process, the app can then be opened. The installation process takes about 1-2 minutes depending on the specs of the smartphone used. The app can be installed on all smartphones running different versions of Android as the platform. Once installed, the icon of the app is selected and the homepage is displayed as shown in Figure 4. The front page displays the information that needs to be filled, and they are: name, identification card number, age and gender. The information is saved together with the 2.4 km run test result conducted. After the information on the front page is completed, users have the opportunity to use the application by running the 2.4 test and get instantaneous result. Users have two options to choose either 'count' or 'manual' by selecting the icon at the bottom of the page. 'Count' serves to start the 2.4 km run test as the user starts running. The result will be displayed automatically. As for 'manual' icon, the users need to enter the duration in the 2.4km run test on their own, only then the test results will be displayed.

Figure 4: The front page of the UPNM Cardio Fitness application

Figure 5: Display of the front page after the 'count' button is selected

Figure 5 shows the front page when 'count' button is selected. The display shows a digital stopwatch that will be automatically activated for the 2.4 km run test carried out by the end user. There are 'start', 'stop', and 'reset' buttons. Users may follow the instructions displayed on the page. 'Result' button is selected to get the run test result immediately. The result is displayed according to the time recorded. A 'menu' button is provided at the bottom of the page to return to the main menu.

The user selects the 'manual' button to know his or her level of achievement or performance in the run test. The image displayed after the button is selected is as shown in Figure 6. Users only need to enter the duration of the run according to the formats used for time. The units according to the formats used for time are minutes, seconds and tenth of a second. Then, the 'result' button is clicked to get the run test result and the level of achievement.

Figure 6: Front page after 'manual' button is selected

Figure 7: Results of 2.4 km test run

The displayed 'result' page shown in Figure 7 shows the result in the 2.4 km run test by a user. Users get an immediate result that will be displayed according to the fitness norms uploaded in the application. The cardiovascular endurance fitness norm of an 18 years old male is used as the database in this application. The page shows individual details filled at the home page of the app. The additional feature available on this app is that each recorded time using this app can be stored in the cloud using Google Fusion Table. A 'save' button is provided for storing data in Google Fusion Table. Figure 7. 'Result' page displayed after run test is conducted either automatically or manually

When the 'save' button is selected, the data is stored immediately into the Google account via Google Fusion Table. Google Fusion Table is an additional feature to manage data to be stored virtually or online [25]. Google Fusion Table was launched in June 2009 by Google Inc. The upside of this function is that data can be accessed anywhere by using internet and a Google account. This method makes it easier for researchers to access data and conduct studies for the next data collection.

Figure 8: Page displayed when 'save' button is selected

5. Discussion and Implication

The result of this research almost resembles the result that was conducted in the study by [14]. Many thought that physical activity and physical fitness are directly related. However, they are defined differently and used in a different context. Physical activities can contribute to physical fitness. Discipline and methods of training must be strictly followed to produce expected results to achieve physical fitness by carrying out physical activities.

The second objective is to build a cardiovascular endurance fitness norm based on the 2.4 km run battery test. The norms were produced from the test conducted on a total of 120 male cadets. The test result is then analysed to construct a fitness norm. However, the norm is limited to male individuals aged 18 years old only. Data mining and production of norms with age ranging from 17 to 25 years old are needed so that the norms can be compared in the future research.

The third objective is to build a mobile application through Android system based on the cardiovascular endurance fitness norm produced. The app, called the UPNM Cardio Fitness was successfully produced using the fitness norms as the baseline of the product. The production has created a new dimension of storing and managing information from the transformation of a test result that can now be easily obtained and accessed. The views and ideas from a committed technology expert has helped the researcher in making sure the goals are achieved. The additional functions in this app can be updated and enhanced from time to time as there are many elements of fitness from other sports that can be included into the application. The new app can be shared easily and is user-friendly.

6. Conclusion

Overall, the product of this research which is the cardiovascular endurance fitness norms for 18 years old UPNM male Military cadets Officer can be used as a guideline to measure the overall level of fitness for the reference of many people. With the production of the norms, the level of cardiovascular fitness norm for an 18 year old male can be assessed through the 2.4 km Run Test. The norms measures and determines the level of fitness accurately when the test is done. Generally, the fitness norms benefit a lot of people especially teenagers who fall into the age category of interest. Amongst the benefits are not only to know the level of fitness but also acts as the motivation to create a better lifestyle that is healthy and safe. Recent studies of Kassim, M. et al. (2015), states that physical fitness is when a person has the ability to perform daily tasks without fatigue with minimal energy consumption and has surplus energy to be used for more challenging activities.

Hopefully, this research gives an impact to several parties which are involved in the management of sports team particularly for trainers to harness the privilege of using technology in recruiting the talents especially in sports. Therefore, it is crucial to have full understanding on the efficient training system to produce quality and planned goals for any team or athletes. Consistent with the research conducted by [26] that proposed an efficient training system using fitness test procedure, it should have high potential to make it as an interesting experience to attract people involved especially youths. In addition to the technological aspect that gained attention especially the millennials who want to obtain information easily without depriving the quality of the information. The outcome of this research could be patented as the proprietary of the UPNM and also utilized as a fitness guideline to produce fitness norms for every age and gender in Malaysia.

References

[1] Sharples, M. 2000. The Design of Personel Mobile Technologies for Lifelong Learning. *Computers and Education*, (34) 177-193.

[2] Salman, A. 2010. ICT, the New Media (Internet) and Development: Malaysian Experience. 15(1. *The Innovation Journal: The Public Sector Innovation Journal,*, 15(1).

[3] Cardinal, D. 2013. *Beyond megapixels: The future evolution of smartphone cameras*.Retrievedfromhttp://www.extremetech.com/extreme: http://www.extremetech.com/extreme/151334-beyond-megapixels-the-future-evolution-of-smartphone-cameras

[4] Taylor, M. 2014. *The Evolution of the Smartphone*. Retrieved from pocketnow: http://pocketnow.com/2014/07/28/the-evolution-of-the-smartphone

[5] eMarketer. 2014. *2 Billion Consumers Worldwide to Get Smart(phones) by 2016*. Retrieved from eMarketer:http://www.emarketer.com/Article/2-Billion-Consumers-Worldwide-Smartphones-by-2016/1011694

[6] Kassim , M., & Ali , N. R. 2015. An Effective Coaching Through 'Coaching Model'. *Journal of Scientific Research & Development, 2*(9), ISSN 1115-7569.

[7] Shanmugapriya, M., & Tamilarasia, A. 2011. Designing an m-learning application for ubiquitous learning environment in the android based mobile devices using web services. *Indian Journal of Computer Science and Engineering (IJCSE)*, 22-30.

[8] Sarwar, M., & Soomro, T. R. 2013. Impact of Smartphone's on Society. *European Journal of Scientific Research, Vol.98 No 2*, 216-226.

IOP Conf. Series: Journal of Physics: Conf. Series **1020** (2018) 012008 doi:10.1088/1742-6596/1020/1/012008

[9] Haron , Z., & Husain, S. 2010. Persepsi Mahasiswa SPI Terhadap Amalan Gaya Hidup Sihat Menurut Islam Dalam Aspek Penjagaan Kesihatan Dan Pemakanan. 1-8. Retrieved from http://eprints.utm.my/10789/1/Persepsi_Mahasiswa_SPI_Terhadap_Amalan_Gaya_Hidup_Sihat_Menurut_Islam_Dalam_Aspek_Penjagaan_Kesihatan_Dan_Pemakanan.pdf

Physical Fitness Index for Assess Fitness Speed Among Army Reserve Officer Training Unit Cadet in Malaysia

Shahrulfadly Rustam[1] and Mohar Kassim [2]

[1]Centre for Coaching Science, National Defence University of Malaysia, Aras 2, Bangunan PTMK, Kem Sungai Besi 57000 Kuala Lumpur, Malaysia

[2]Centre for Coaching Science, National Defence University of Malaysia, Aras 2, Bangunan PTMK, Kem Sungai Besi 57000 Kuala Lumpur, Malaysia

Email: shahrulfadly@upnm.edu.my

Abstract. The aim of this study is to develop physical fitness index (PFI) for physical fitness among Army Reserve Officer Training Unit Cadet Malaysia. This study use 30 meter speed run as a physical fitness test battery to develop physical fitness index (PFI) and to evaluate the subject fitness speed. 212 male respondent (N=212) was selected in this study including Army Reserve Officer Training Unit Cadet of National Defence University of Malaysia. 30 meter sprint was used as a instrument for this study. The methodology will be adopted for this study is quantitative research in the form of a quasi-experiment. Quasi-experimental methods is used to measure and evaluate the level of physical fitness and develop Physical Fitness Index especially in speed. The design of this study is quasi-experimental study design with pre-test and post-test. The study design is quasi-experimental research design in which the data is obtained through the practical test in the field. The data were analysed by using the SPSS software version 20 to calculate the mean, standard deviation and t-test for develop physical fitness index (PFI) and to evaluate the fitness speed level for Army Reserve Officers Training Unit Cadet Malaysia. The findings showed mean and standard deviation for develope physical fitness index is (M=4.84) and (SD=0.48). The t-test for evaluate fitness level in speed for pre-test and post-test is significantly difference ($p \leq 0.05$). The implication at this study is that the develope of standard physical fitness index is able to identify the level of physical fitness among Army Reserve Officer Training Unit Cadet Malaysia.

Keywords: Physical fitness, Physical fitness index, test battery, Army Reserve Officer Training Unit Cadet, 30 meter sprint.

1. INTRODUCTION

Physical fitness is a foundation of holistic wellness. Physical fitness concepts include of physical, mental, emotional, social and spiritual aspects. Physical fitness is a universal concept that requires an individual to be totally healthy and strong. The life in this modern age, physical fitness is very important for a person.

The perception of which only individuals who play sports only require physical fitness should be denied [2]. The important thing to be taken into account by each individual is deteriorating physical condition will impact on the capacity of individual mental and social aspects. It also meant, in other words, fitness is at the minimum level will influence and affect mental function and ability to react with other things.

Content from this work may be used under the terms of the Creative Commons Attribution 3.0 licence. Any further distribution of this work must maintain attribution to the author(s) and the title of the work, journal citation and DOI.
Published under licence by IOP Publishing Ltd

Fitness is the ability to compete with a balanced and complete life [7]. But it also involves the whole of the social, psychological fitness and physiology and fitness is determined by environment, heredity, genes and lifestyle [7].

It is generally realized that armed forced require a specific level of by and large or finish the fitness to meet the physical requests of war. Hopping, slithering, moving, ceasing, beginning, jumping, climbing, pushing, sprinting from cover to cover, conveying overwhelming burdens long separations, and as yet having the capacity to finish the current mission speak to a short rundown of the required undertakings put upon a armed forced. The main key measurable fitness components include mobility, strength, endurance and flexibility [4].

The human element and physical fitness of the individuals from military have dependably been and will stay a standout amongst the most imperative objectives and characteristics of the national safeguard in the military around the world, in any case superior and improvement of weapons innovation [5] [6]. Past involvement in history proposes that physical preparing was of awesome significance, through preparing of both troopers and officers for fruitful summon and control.
Physical fitness is an important marker of the health of the armed forces and security [3]. The security forces should have a high level of physical fitness due to the exercise of physical activity performed during military exercises and in war [8]. Monitoring of the physical fitness level of the armed forces and security are important from a performance point of view, and also to assess their combat capability. Kassim & Mokhtar (2016) stated that the important of coaches requiring knowledge in the coaching process are important to build up the quality of fitness level using the norms of physical fitness.

Speed is one of the physical fitness elements. It's involves the muscles of the lower body of the human body. Speed is defined as the capacity and capability of moving a body part or the entire body with the supreme and utmost likely speed or velocity [11]. Speed over various distance is an important quality in many sports. There are many instances in team or racquet sports where the demands of speed are characterised more by rapid changes of direction rather than running in a straight line [12] [13]. These changed of direction movements, often describe as agility, may be initiated to either evade or pursue an opponent, or to react to a moving ball. When reacting, it has been shown that the nature of the stimulus (timing and location) influences performance in agility tasks, and therefore perceptual factors are significant for agility performance [14] [15].

2. PROBLEM STATEMENT

Physical fitness index is a criterion that can be used to help interpret the physical fitness level of a person in a sports skills [2]. Physical Fitness index can help coaches learn military physical performance levels of physical fitness army officers who participated in physical activities. Through Fitness index also, army physical trainer can identify the strengths and weaknesses of military officers who served in the armed forces of Malaysia.

Knowledge of the physical fitness level of military officers to help army physical fitness trainers provide appropriate military training to increase the level of physical fitness for a military officer. Developing Physical Fitness index among military officers in Malaysia indispensable at present for view and maintain the performance level of physical fitness and military officials in Malaysia.

Difficulty obtaining a Physical Fitness index lead our country still using tests that do not have a standard procedure for measuring the performance of physical fitness. This can cause major problems to the Physical Trainer of the armed forces. So far, no Physical Fitness index can be adopted for determining the level of physical fitness achievement army officer in Malaysia. So far, don't have any test procedures and standard physical fitness index that can be introduced and implemented in the armed forces of Malaysia

So far, the military officers who underwent physical training or physical fitness cannot know the performance level of physical fitness because their performance is not measured and assessed systematically. Physical trainer of the armed forces are not successful using a consistent measurement tests because there is no Physical Fitness standard index, therefore achievement test scores were not able to demonstrate the level of physical fitness to the actual military officers.

Thus, the researchers plan to investigate the level of physical fitness among the Army Reserve Officer Training Unit (ROTU) Cadet in Malaysia through the development of a physical fitness standard index for Army ROTU Cadet.

3. THE AIM OF RESEARCH

This paper conducted research to develop physical fitness index for fitness speed and evaluate the fitness speed level among Army Reserve Officer Training Unit (ROTU) Cadet in Malaysia.

4. METHOD OF RESEARCH

This study conducted a quantitative research in the form of a quasi-experiment. Quasi-experimental methods is used to measure and evaluate the level of physical fitness and develop Physical Fitness index. This experimental observation study use a speed physical fitness test battery that best suits the activity.

The design of this study is quasi-experimental study design with pre-test and post-test. The study design is quasi-experimental research design in which the data is obtained through the practical test in the field [1].

Population that used to conduct this research is composed of Army Reserve Officer Training Unit (ROTU) Cadet aged 19 to 21 years. The sample was made up of Army Reserve Officer Training Unit (ROTU) of National Defence University of Malaysia aged 19 to 21 years. A total 212 male Army Reserve Officer Training cadet were selected as a study subjects. The purpose of this study is to develop a physical fitness index of fitness speed by using 30 meter sprint test battery for Army Reserve Officer Training Unit Cadet, National Defence University of Malaysia. The sample size needed for the creation of physical fitness index should be at least 200 people for each variable [9].

Table 1: Selected test battery for study instrument

Physical Fitness Component	**Test battery**
Speed	30 meter sprint

A pilot study was conducted to determine the reliability and validity of the instruments to be used in the study. A pilot study is a small study or trial version that has been done in preparation for a larger study or major. The main purpose of this pilot study was to determine the validity and reliability of the instrument. This pilot study is also a pre-test to try a special instrument.

This research will conduct a pilot study to test the validity and reliability of the physical fitness test battery selected for forming the standart physical fitness norm standards for officers of the Army Reserve Officers Training Unit Cadet Malaysia. This study use the test repetition to get the coefficient of reliability. A total of 50 military cadets Officer National Defence University of Malaysia who are not involved with the study respondents used as a pilot study. In this study, the objective of the pilot study is to ensure the validity and reliability of the battery of tests chosen to form indices standart physical fitness index of Malaysian armed forces.

IOP Conf. Series: Journal of Physics: Conf. Series **1020** (2018) 012009 doi:10.1088/1742-6596/1020/1/012009

Table 2: Method of data analysis

Part A	Analysis type	Measurement tool
A	Background data of respondent	Frequency, percentage
B	Develop physical fitness index for speed	Mean, standard deviation
C	Pre-test and post-test fitness level for speed	Paired sample T-test

5. RESULT AND DISCUSSION

All raw data that will be gathered will be collected and analyzed by the scores of selected physical fitness test battery by using SPSS version 20. The raw data that be analyzed by the scores of physical fitness test batteries to get the mean and standard deviation for develope speed physical fitness index and t-test for evalute pre test and post test speed fitness level for Army Reserve Officers Training Unit Cadet Malaysia. This study are used a battery of physical fitness test 30 meter speed for speed fitness.

Data analysis are involve several steps of storing data, not code, isolate the data, make calculations and analyse data. All the measurements of the test battery are collected and recorded. Revisions to data of fitness tests will be done to ensure that the data is complete and in accordance with established procedures. Structural of data in this study was a score and the data is a quantitative design.

This normative research will be used to develop physical fitness index through a selected test battery and statistical inference will be used to analyse the level of physical fitness according to age and gender. Paired sample T-test are used to see the difference between physical fitness in pre-test and post-test of fitness level of the subject. The achievement level of physical fitness of subjects are analysed by looking at the score data by officers Cadet Force Army Reserve Officers Training Unit Cadet Malaysia during battery tests performed.

For this study, researchers have set three levels of measurement that must be taken by the subject. Firstly is a pre measurement level, the researchers will give a briefing to a selected sample of the purpose of this measurement is carried out.

Secondly is a level measurement, for this stage the sample will be classified into three groups based on age (19 years, 20 years and 21 years) on Army Reserve Officers Training Unit Cadet Malaysia will be selected randomly. With these three groups, the researchers set out two steps to carry out the measurement. First, the researchers will deliver forms to sample score sample which is required to fill in the required information and the second one is a researchers will describe how the test is carried out on a sample, the procedure to be followed and how the measures will taken.

Thirdly is a test procedure will be done by each test battery to be measured according to the components of fitness. After the test battery conducted, researchers analyze the data taken from measurements made during the pre-test and post-test. Researchers will analyze the data to determine the fitness level of the subject.

5.1 *Background data for respondent*

Table 3: Descriptive data for age

Subject	N	Age	Frequency	Percentages	Valid percent	Cumulative percent
Male	212	19	94	44.3	44.3	44.3
		20	75	35.4	35.4	79.7
		21	43	20.3	20.3	100.00

Table 4: Descriptive data for race

Subject	N	Race	Frequency	Percentages	Valid percent	Cumulative percent
Male	212	Malay	189	89.2	89.2	89.2
		Chinese	2	0.9	0.9	90.1
		Indian	19	9.0	9.0	99.1
		Others	2	0.9	0.9	100.00

5.2 *Data for develop physical fitness index for speed*

Table 5: Descriptive statistic on respondent's 30 meter sprint test score

Descriptive statistic	Test score
Minimum	4.00
Maximum	6.73
Mean	4.84
Standard deviation	0.48
Skewness	0.881
N	212

Table 6: Physical Fitness Index for speed

Category	Duration (ss)	Score
Excellent	< 4.11	5
Good	4.12-4.60	4
Fair	4.61-5.08	3
Poor	5.09-5.57	2
Very poor	5.58 <	1

Note: ss = second

Table 6 shows the physical fitness index produced based on the 30 meter sprint test results which the mean and standard deviation is (M=4.84, SD=0.48). There are 5 categories that are classified in the grade production of a physical fitness index which are; Excellent (5), Good (4), Fair (3), Poor (2), and Very Poor (1). Grading a test in the production of physical fitness index enables different levels of potential in a particular group according to the abilities of each one of the individuals. Every category of accomplishment states the scores achieved starting from the highest score which is 5 to 1, which 1 indicates the lowest score [10]. From the table, the highest achievement which is Excellent with the score of 5 starts from 4.11 second and below, the next level is Good with the scores of 4 is from 4.12 to 4.60 second, the Fair level with the score of 3 is from 4.61 to 5.08 second, and the Poor level with the score of 2 is from 5.09 to 5.57 second and lastly, the Very Poor level with the score of 1 is from 5.58 second and above.

5.3 Data for fitness level of speed

Table 7: Frequency and percentages for fitness level of speed among Army Reserve Officers Training Unit Cadet National of Defence University of Malaysia

	Frequency	**Percent**	**Valid percent**	**Cumulative percent**
Excellent	29	13.7	13.7	13.7
Good	84	39.6	39.6	53.3
Fair	80	37.7	37.7	91.0
Poor	17	8.0	8.0	99.1
Very poor	2	0.9	0.9	100.0
Total (N)	212	100.00	100.00	

Table 8: Paired Samples Statistic for pre-test and post-test fitness speed level

		Mean	**N**	**Std. Deviation**	**Std. Error Mean**
Pair 1	Pre-test fitness speed	4.8497	212	.48424	.03326
	Post-test fitness speed	4.5942	212	.39537	.02715

Table 9: Paired samples test for pre-test and post-test fitness speed level

		Paired Differences					**t**	**df**	**Sig.(2-tailed)**
		Mean	Std. Deviation	Std. Error Mean	95% Confidence Interval of the Difference				
					Lower	Upper			
Pair 1	Pre-test fitness speed – Post-test fitness speed	.25552	.50916	.03497	.18659	.32445	7.307	211	.000

Paired sample T-test analysis is used to compared mean score pre-test fitness speed and post-test fitness speed for Army Reserve Officers Training Unit Cadet National of Defence University of Malaysia. The findings showed mean score post-test fitness speed (M=4.59, SD=0.39) is more speed and it's significant compared to mean score pre-test (M=4.84,SD=0.484), t (211)=7.30, $p \leq 0.05$

6. CONCLUSIONS

Physical fitness is generally include physical, mental, emotional, intellectual, spiritual and social aspects. Physical fitness is a comprehensive concept and it allows individuals to live better, energized and confident. Therefore, in this more advanced age, physical fitness is very important for a person's health. Deterioration of the physical condition of the person will give a real impact on individual social mental strength. It also meant, in other words, less active physically can affect mental function and ability to react with other things.

The ability to identify the performance of physical fitness can benefit the individual to choose the type of sport and career interest. Through the results of the test and measurement, allowing coaches and sports science practitioners use physical fitness performance information as a basis for placing groups of individuals based on actual physical fitness level during a training session in progress. Knowledge of the performance of individual physical fitness can help coaches and sports science practitioners to design strategies and methods that are effective in delivering sports training sessions and games.

This study will be done is to establish physical fitness index against the Army Reserve Officer Training Unit Cadet Malaysia accordance with demographic factors in Malaysia. In fact, this study also, is to help army physical trainer assess the level of physical fitness among Army Reserve Officer Training Unit Cadets Malaysia especially in the National Defence Universities of Malaysia. Even with the establishment of a standard physical fitness index, it can improve the quality of physical fitness level among army officer, especially the Cadet Officer of Army Reserve Officer Training Unit. Not only that, the army physical trainer can plan and identify the types of training according to Cadets officer of Army Reserve Officer Training Unit.

ACKNOWLEDGEMENTS

First of all, the authors wish appreciation to the Centre of Coaching Science, and the Research and Innovation Division of the National Defence University of Malaysia for giving the opportunity to conduct and publish the result of this research under National Defence University of Malaysia research grant UPNM/2016/GPJP/4/SS/2. Special thanks go to Cadet Officer of Army Reserve Officer Training Unit, National Defence University of Malaysia for supporting the studies of the authors that enabled them to conduct this research.

REFERENCES

[1] Ahmad H. & Gunathevan. 2015. 900 push-up test norms for sport science students Sultan Idris Education university. *International Journal of Development and Emerging Economics*, 3 1-9.

[2] Ahmad H. 2015. Pengujian pengukuran dan penilaian pendidikan jasmani. (1st). Selangor. Dubook Press Sdn Bhd.

[3] Erikssen, G. 2001. Physical fitness and changes in mortality: the survival of the fittes. *Sport Med.* 31 (2) ,121- 127.

[4] Roy T.C, Springer B.A, McNulty V & Butler N.L. 2010. Physical fitness. *Mil Med.* 175(8), 14–20.

[5] Rodic N. 1993. The influence of basic training on physical fitness. Belgrade: Novi glasnik.

[6] Rodic N. 1994. A new program for physical training of soldiers. Belgrade: Novi glasnik..

[7] Kassim, M. 2016. Konsep Latihan Dalam Sukan. Selangor: Pekan Ilmu Publications Sdn.Bhd.

[8] Kassim , M., & Mokhtar , R. S. 2016. The Evaluation of Cardiovascular Endurance Test for Male Cadet Officer. *Modern Applied Science, 10*(2), 11-16.

[9] Knapik, J., Daniels, W., Murphy, M.,Fitzgerald, P., Drews, F., & Vogel, J. 1990. Physiological factors in infantry operations. *Eur J Appl Physio Occup Physio* 60 (3), 223-238.

[10] Morrow, J,R, Jackson, A.W., Dish, J.G, & Mood, D.P. 2005. Measurement and evaluation in human performace (3rd). Champaign IL: Human Kinetic.

[11] Ahmad, H. 2004. Pengukuran Kecergasan Motor. Tanjung Malim: Quantum Books.

[12] Abdul L.S. 2017. Impact of speed training and plyometric exercise on triple jump performance among fewer than 17 boys. *International Journal of physical education, fitness and sports*, 46-53

[13] Twist P.W, Benicky D. 1996. Conditioning lateral for multi-sport athletes: practical strength and quickness drills. *Strength Cond.* 19 (3), 10-9

[14] Gambetta V. 1996. How to develop sport specific speed. *Sports Coach* 19 (3), 4-22

[15] Chelladurai P. 1976. Manifestation of agility. *CAHPER J* 42(3), 36-41

[16] Chelladurai P, Yuhasz M.S., Sipura R. 1977. The reactive agility test. *Percept Mot Skills* 44(13), 19-24

IOP Conf. Series: Journal of Physics: Conf. Series **1020** (2018) 012010 doi:10.1088/1742-6596/1020/1/012010

Effects of Aqua Exercises Towards Improving The Quality of Life (QoL) of Obese Women in Malaysia

Noor Liyana binti Karim[1], Asma Diyana binti Abd Jalil[1], Noor Haninah binti Hasri[1], Hezlin Aryani binti Abd Rahman[1], Maisarah binti Shari[2] and Nur Izzati binti Idris[2]

[1]Centre of Statistical and Decision Science Studies, Faculty of Computer and Mathematical Sciences, Universiti Teknologi Mara Malaysia 40450 Shah Alam

[2]Faculty of Sports & Recreational Science, Universiti Teknologi Mara Malaysia 40450 Shah Alam

Corresponding address : nliyanakarim@gmail.com, asmadiyana43@gmail.com, noorhaninahhasri@gmail.com, hezlin@tmsk.uitm.edu.my, izzatiidris12@gmail.com

Abstract. Aqua exercise is a form of water exercise, done in the water which is beneficial for weight loss as well as improving the quality of life. It is suitable for all age group and fitness levels whereby due to the water buoyancy makes it easier to perform exercises especially for obese and knee-injured people. However, there was not much study done to measure the effectiveness of the aqua exercises in improving the quality life. Thus, this study aims to investigate and compare the effectiveness of aqua exercises towards obese women within eight domains of the Quality of Life (QoL). This study uses the 36-Items Short Form Health Survey (SF-36) questionnaire and a purposive sample of 61 participants to measure the effectiveness of the aqua exercise before and after 36 days of aqua workout. As the nature of the data collected was not normally distributed, hence the Wilcoxon signed rank test was used as the statistical method of analysis. The findings of this study showed that there was a significant difference between the overall QoL pre and post since the p-value < 0.05. In addition, it was also found that five out of the eight domains of QoL; the physical functioning, general health, social functioning, mental health, and health transition were the domains showing significant difference between the pre and post-test (p-value < 0.05), and where majority of the participants showed a significant improvement after the aqua workouts. Thus, it can be concluded that aqua exercises is effective in improving the general QoL of obese women.

1. Introduction

Obesity is categorized as an abnormal or excessive fat accumulation in the body that may effects the health of obese people. It can be measured by measuring the Body Mass Index (BMI) which is a simple index of weight for height that commonly used to identify underweight, normal, overweight and obesity in adults. BMI can be calculated by dividing a person's weight in kilograms with the square of his height in meters (kg/m^2). The people with BMI greater than or equal to 30 are categorized as obese.

The current statistics by World Health Organization reported that over 600 million adults were obese and about 11% of men and 15% of women of the world's adult population were obese up till 2014. The prevalence of obesity had increased rapidly in Malaysia in the last decade of the twentieth century covering till the years of 2015.

Content from this work may be used under the terms of the Creative Commons Attribution 3.0 licence. Any further distribution of this work must maintain attribution to the author(s) and the title of the work, journal citation and DOI.
Published under licence by IOP Publishing Ltd

Malaysia is one of the ASEAN country that have a high prevalence of obesity. Malaysia is in the ranked of sixth in the Asia-Pacific region for obesity and tops the list in South-East Asia for both obesity and diabetes. The National Health & Morbidity Survey (NHMS) 2015 reported that the national prevalence of overweight, obesity and abdominal obesity rose by 0.6%, 2.6% and 2.0% respectively as compared to the previous findings of NHMS 2011.

Obesity is known as a major determinant of chronic terminal diseases such as cancers, gall bladder diseases, respiratory problems, musculoskeletal disorders and induces type 2 diabetes mellitus (Ghee, 2016). It appears to be more prevalent in women compared to men since women are less likely to do exercises during their leisure time when compared to men. Most of the previous studies also reported that women are more obese than men (Ismail et al., 2002; Ghee, 2016).

One way of reducing the percentage of obesity in Malaysia is by encouraging these obese people to exercise. There are two types of exercises that can be practice which are land-based exercises and water-based exercise. There are many types of water-based exercises such as aqua zumba, water yoga, aqua aerobics, and aqua jogging. The practice of exercises in water is suitable for obese people since it can increase energy consumption compared to running training on land, in addition reducing the risk of injury. In addition, usually the obese people will have certain knee-injury problem. This adds to the need of a low-knee-impact exercise to reduce the weight problem and aqua exercising is a great option for these people.

Previous studies stated that water-based exercises are more effective compared to land exercises especially for obese and older people with knee and joint problems. This is because 90% water buoyancy assist in supporting the body and making exercise feel easier since it acts as a shock absorb and reducing stress on joints. It gives the obese people to do exercises without feeling pain.

Since women are more prevalent in becoming obese as compared to men and achieving an ideal weight through exercise is important in improving the quality of life of obese women, hence, this study purpose to investigate the effectiveness of aqua exercises (aqua zumba and aqua jogging) towards obese women and to compare the effectiveness of aqua exercises towards obese women within the eight domains of Quality of Life (QoL).

2. Exercising and Quality of Life (QoL)

Exercise is a vital part in reducing the weight. Exercise refers as any body movement that enhances health and increases energy expenditure. Exercises can be divided into two which are land exercise and water exercise. Both exercises give benefits towards body and lifestyle. Land exercise is the familiar exercise such as jogging, running, aerobic, zumba and many more. Oh et al., (2015) state that land based exercises improve the lower limb strength of elderly people, as well as other physical functions. Land exercise offers tremendous benefits, but too often benefits are accompanied by aches and pain, overheating, sweating and feeling exhaustion (Madhankumar & Sundar, 2016). Aqua exercises also known as water aerobics, aqua fitness or aqua fit. Water aerobics is a form of aerobic activity involving long dynamic exercises performed at moderate intensity (Kantyka, Herman, Roczniok, & Kuba, 2015). There are different forms of aqua exercises such as aqua zumba, water yoga, aqua aerobics, and aqua jog (Madhankumar & Sundar, 2016). Pasetti, Gonçalves & Padovani (2012) cited that the practice of exercises in water environments may be interesting for obese people since it can increase energy consumption compared to running training on land and it can reduce the risk of injury.

Improving ones' quality of life in the recent years has been an important study. Quality of life (QoL) is defined as how an individual measures the goodness of multiple aspects of their life (Theofilou, 2013). QoL are subjective experiences, thus, making effective measurement difficult. Knowing the patients' QoL are vital to better understanding how a disease, in our case obesity, affects the patients' day to day life and the personal burden resulting from the illness (New & Somani, 2016).

According to Oh et al. (2015) water-based exercise gives greater effect in improving daily-life-activities related and other items, such as role-physical and bodily pain. Wouters, Van Nunen, Geenen, Kolotkin, & Vingerhoets (2010), also reported that aqua jogging give effects towards the QoL of obese women since there are improvements in QoL. Studies done proved that only some aspects of QoL were found to change as the result of a 6-month period of an exercise program, as the program was not long enough to obtain maximum results. However, visible changes of overall QoL can still be achieved (Vécseyné Kovách, Kopkáné Plachy, Bognár, Olvasztóné Balogh, & Barthalos, 2013).

3. Methodology

A sample of 61 participants was selected using purposive sampling strategy. The purposive sampling technique is a non-probabilistic sampling technique, where some criterions were imposed in the selection of the sample. The criterions includes selecting participants were female aged 20 to 59 years old, having body mass index (BMI) and fat percentage range from 30 to 40 kg/m2 and 35 to 45% respectively. A total of 110 participants were participated in this study and randomly assigned into 3 different intervention groups (aqua Zumba, aqua jogging and control). However, due to some constraints only 61 participant were able to complete this experiment within the time given (36 days).

This study measures the effectiveness of aqua exercise in improving the QoL by using the 36-Items Short Form Health Survey (SF-36), which divided the QoL into eight domains of QoL; (i)physical functioning, (ii)role functioning, (iii)bodily pain, (iv)general health, (v)vitality, (vi)social functioning, (vii)mental health, and (viii)health transition. The data were scored by using the scoring rules for the RAND 36-Item Health Survey.

The statistical method use in this study was Wilcoxon Sign Rank Test. This method was selected because the data was not normally distributed.

4. Results and Discussions

4.1 Description of Data

The participants were assigned with of two different types of aqua exercises which are aqua jogging and aqua Zumba. There was also one control group. The highest respondents come from aqua jogging with 24 (39.3%) respondents while aqua zumba with 22 (36.1%) respondents. The least of our respondents were control group with 15 (24.6%). Due to the experimental nature of this study, the number of respondents who completed the whole study was very limited.

Table 1. Descriptive Statistics showings the respondents of Aqua Exercises.

Variable	n (%)	Mean (SD[a])
Age		30.79 (10.52)
Pre Weight		82.70 (12.67)
Post Weight		81.08 (12.64)
Height		158.54 (5.77)
Pre BMI		32.86 (4.25)
Post BMI		32.16 (4.35)
Types of Aqua Exercise		
Jogging	24 (39.3)	
Zumba	22 (36.1)	
Control	15 (24.6)	

[a]*Standard Deviation*

4.2 Testing Overall Score of QoL Before and After Workout

To test the difference between overall QoL for the pre-workout and post workout, we used the Wilcoxon signed rank test since the data was not normally distributed. The table 2 below shows that the highest score of overall total post – overall total pre were positive ranks with total of 39 respondents. This value means that majority (39 respondents) have the score of overall total post higher than the score of overall total pre. The mean value of positive and negative rank was 34.06 and 20.13 respectively. Whereas the significance level shown in table 3 concluded that the score between overall QoL for pre-workout and post-workout experiments was statistically significant since the p-value (0.000) < 0.05. Therefore, we can conclude that the two overall scores of QoL, for the group before and after joining the aqua workouts were significantly different and that the majority of the participants achieved a generally improved QoL after joining the 36-days aqua exercise program.

Table 2. The Negative and Positive Ranks Value of Overall Pre and Post.

		n	Mean Rank
Overall Total Post – Overall Total Pre	**Negative Ranks**	19[a]	20.13
	Positive Ranks	39[b]	34.06
	Ties	3[c]	

[a] *Overall Total Post < Overall Total Pre* [b] *Overall Total Post > Overall Total Pre* [c] *Overall Total Post = Overall Total Pre*

Table 3. The Differences between Overall QoL Score for Pre and Post Workout.

	Z-Value	P-Value
Overall Total Post – Overall Total Pre	-3.662	0.000

Table 4. The Negative and Positive Ranks Value of Total Pre and Post Score within Domains of QoL.

		n	Mean Rank
Total Post PF – Total Pre PF	**Negative Ranks**	15[a]	16.83
	Positive Ranks	37[b]	30.42
	Ties	9[c]	
Total Post RF – Total Pre RF	**Negative Ranks**	18[d]	24.03
	Positive Ranks	24[e]	19.60
	Ties	19[f]	
Total Post BP – Total Pre BP	**Negative Ranks**	13[g]	16.85
	Positive Ranks	18[h]	15.39
	Ties	30[i]	
Total Post GH – Total Pre GH	**Negative Ranks**	8[j]	22.38
	Positive Ranks	42[k]	26.10
	Ties	11[l]	
Total Post Vitality – Total Pre Vitality	**Negative Ranks**	19[m]	18.37
	Positive Ranks	20[n]	21.55
	Ties	22[o]	
Total Post SF – Total Pre SF	**Negative Ranks**	15[p]	24.60
	Positive Ranks	32[q]	23.72
	Ties	14[r]	
Total Post MH – Total Pre MH	**Negative Ranks**	14[s]	20.36
	Positive Ranks	28[t]	22.07
	Ties	19[u]	
Total Post HT – Total Pre HT	**Negative Ranks**	10[v]	25.10
	Positive Ranks	36[w]	23.06
	Ties	15[x]	

[a] Total Post PF < Total Pre PF [b] Total Post PF > Total Pre PF [c] Total Post PF = Total Pre PF [d] Total Post RF < Total Pre RF [e] Total Post RF > Total Pre RF [f] Total Post RF = Total Pre RF [g] Total Post BP < Total Pre BP [h] Total Post BP > Total Pre BP [i] Total Post BP = Total Pre BP [j] Total Post GH < Total Pre GH [k] Total Post GH > Total Pre GH [l] Total Post GH = Total Pre GH [m] Total Post Vitality < Total Pre Vitality [n] Total Post Vitality > Total Pre Vitality [o] Total Post Vitality = Total Pre Vitality [p] Total Post SF < Total Pre SF [q] Total Post SF > Total Pre SF [r] Total Post SF = Total Pre SF [s] Total Post MH < Total Pre MH [t] Total Post MH > Total Pre MH [u] Total Post MH = Total Pre MH [v] Total Post HT < Total Pre HT [w] Total Post HT > Total Pre HT [x] Total Post HT = Total Pre HT

3.3 Testing the 8 Domains of QoL Before And After Workout

To test the differences between the 8 domains of QoL for pre-workout and post-workout, the Wilcoxon signed rank test was used since each of the domains tested were not normally distributed. Table 4 shows 5 out of 8 domains that have the highest positive ranks value where the total post greater than total pre score within domains of QoL were physical functioning with respondents of 37 (30.42), role functioning with respondents of 24 (19.60), general health with respondents of 42 (26.10), social functioning with respondents of 32 (23.72), mental health with respondents of 28 (22.07) and health transition with respondents of 36 (23.06). The remaining domains were bodily pain and vitality has the same score between total post and pre with respondents of 30 and 22 respectively.

Based on the table 5, only 5 domains were significantly difference for the pre-workout and post-workout, which are; physical functioning, general health, social functioning, mental health and health transition (p-value < 0.05). The remaining 3 domains (role functioning, bodily pain and vitality) were not found to be significantly different (p-value > 0.05).

Therefore, it can be concluded that majority of the participants have felt improvements in 5 out of 8 domains of QoL, which are: physical functioning, general health, social functioning, mental health and health transition. However, in 3 out of 8 domains, the participants felt no improvement or lacking of improvements. The three domains are; role functioning, bodily pain and vitality.

Table 5. The Difference between Total Pre and Post Score Within Domains of QoL.

Total Post –Total Pre	Z-Value	P-Value
Physical Functioning (PF)	**-4.001**	**0.000**
Role Functioning (RF)	-0.239	0.811
Bodily Pain (BP)	-0.571	0.568
General Health (GH)	**-4.462**	**0.000**
Vitality	-0.580	0.562
Social Functioning (SF)	**-2.101**	**0.036**
Mental Health (MH)	**-2.103**	**0.035**
Health Transition (HT)	**-3.170**	**0.002**

5. Conclusion

This study was to investigate the effectiveness of aqua exercises towards improving the QoL of obese women within the eight domains of QoL. The results shows that overall, the aqua exercise improved the QoL of the 61 obese participants significantly. There were also significant improvements found in the overall scores for 5 out of 8 domains of QoL exercises. Hence, the aqua exercises were effective in improving the overall QoL of obese women.

The five domains of QoL which were found to improve significantly are the physical functioning, general health, social functioning, mental health, and health transition. However, the other three domains of QoL which are role functioning, bodily pain and vitality were found not to have significant improvements in the QoL of obese women.

As stated in a study by Vécseyné Kovách, Kopkáné Plachy, Bognár, Olvasztóné Balogh, & Barthalos (2013), only a few domains will be significantly showing improvements in the QoL due to many unforeseen reasons. Among the biggest reasons are the duration of the experiment (36 days) were not enough to study improvements in a person's body and the small sample size. Although the program started of with quite a number of participants, however, due to the human nature, it can't be avoided to have a portion of drop-outs throughout the experimentation period.

6. Recommendation

For the future study, we recommend to investigate the effectiveness regarding the types of aqua exercises (aqua Zumba & aqua jogging) in further investigating the important domains in improving the QoL of obese women. It is also important to have a bigger sample size for future studies.

7. Acknowledgement

We would like to acknowledge the Universiti Teknologi Mara Shah Alam and Studies Faculty of Computer and Mathematical Sciences for giving opportunities to us to carry out this research. We would also acknowledge to our advisor Puan Hezlin Aryani binti Abd Rahman, Ms Maisarah binti Shari and other lecturers for giving advice and support to finish this research.

References

[1] Ismail, M. N., Chee, S. S., Nawawi, H., Yusoff, K., Lim, T. O., & James, W. P. T. (2002). Obesity in Malaysia. *Obesity Reviews*, *3*(3), 203–208. https://doi.org/10.1046/j.1467-789X.2002.00074.x.

[2] Kantyka, J., Herman, D., Roczniok, R., & Kuba, L. (2015). Effects of Aqua Aerobics on Body Composition, Body Mass, Lipid Profile, and Blood Count in Middle-Aged Sedentary Women. *Human Movement*, *16*(1), 9–14. https://doi.org/10.1515/humo-2015-0020.

[3] Lim, K. G. (2016). A review of adult obesity research in Malaysia. *Medical Journal of Malaysia*, *71*, 1–19. https://doi.org/10.4103/0022.

[4] Madhankumar, M., & Sundar, D. M. (2016). Effect of aqua aerobic exercises and aerobic exercises on body mass index parameters among college men students. *International Journal of Innovative Knowledge Concepts*, *2*(9). Retrieved from http://www.ijikc.co.in/sites/ijikc/index.php/ijikc/article/view/259.

[5] Ministry of Health Malaysia. The National Health & Morbidity Survey (NHMS) (2015). *VOLUME II : Non-Communicable Diseases, Risk Factors & Other Health Problems.*

[6] New, F., & Somani, B. K. (2016). A Complete World Literature Review of Quality of Life (QOL) in Patients with Kidney Stone Disease (KSD). *Current Urology Reports*, *17*(88), 1–6. https://doi.org/10.1007/s11934-016-0647-6.

[7] Oh, S., Lim, J.-M., Kim, Y., Kim, M., Song, W., & Yoon, B. (2015). Comparison of the effects of water- and land-based exercises on the physical function and quality of life in community-dwelling elderly people with history of falling: A single-blind, randomized controlled trial. *Archives of Gerontology and Geriatrics*, *60*(2), 288–293. https://doi.org/10.1016/j.archger.2014.11.001.

[8] Pasetti, S. R., Gonçalves, A., & Padovani, C. R. (2012). Continuous training versus interval training in deep water running: health effects for obese women. *Revista Andaluza de Medicina Del Deporte*, *5*(1), 3–7. https://doi.org/10.1016/S1888-7546(12)70002-3.

[9] Rand Corporation (n.d). 36-Item Short Form Survey (Sf-36) Scoring Instructions. Retrieved from https://www.rand.org/health/surveys_tools/mos/36-item-short-form/scoring.html

[10] Theofilou, P. (2013). Quality of life: Definition and measurement. *Europe's Journal of Psychology*, *9*(1), 150–162. https://doi.org/10.5964/ejop.v9i1.337.

[11] Vécseyné Kovách, M., Kopkáné Plachy, J., Bognár, J., Olvasztóné Balogh, Z., & Barthalos, I. (2013). Effects of Pilates and aqua fitness training on older adults' physical functioning and quality of life. *Biomedical Human Kinetics*, *5*(1), 22–27. https://doi.org/10.2478/bhk-2013-0005.

[12] World Health Organization (2016). Obesity and overweight. Retrieved from http://www.who.int/mediacentre/factsheets/fs311/en/

[13] Wouters, E. J. M., Van Nunen, A. M. A., Geenen, R., Kolotkin, R. L., & Vingerhoets, A. J. J. M. (2010). Effects of Aquajogging in Obese Adults: A pilot study. *Journal of Obesity*, *2010*. https://doi.org/10.1155/2010/231074.

IOP Conf. Series: Journal of Physics: Conf. Series **1020** (2018) 012011 doi:10.1088/1742-6596/1020/1/012011

Automatic Line Calling Badminton System

Syahrul Affandi Saidi[1*], Nurabeahtul Adawiyah Zulkiplee,[1] Nazmizan Muhammad[1] , Mohd Sharizan Md Sarip[1]

[1]Faculty of Engineering Technology, Universiti Malaysia Perlis, Kampus UNICITI Alam, Sungai Chuchuh, Padang Besar, 02100, Perlis, Malaysia

syahrulaffandi@unimap.edu.my

Abstract. A system and relevant method are described to detect whether a projectile impact occurs on one side of a boundary line or the other. The system employs the use of force sensing resistor-based sensors that may be designed in segments or assemblies and linked to a mechanism with a display. An impact classification system is provided for distinguishing between various events, including a footstep, ball impact and tennis racquet contact. A sensor monitoring system is provided for determining the condition of sensors and providing an error indication if sensor problems exist. A service detection system is provided when the system is used for tennis that permits activation of selected groups of sensors and deactivation of others.

1. Introduction

Badminton is one of the favorites sports and pastimes for most Malaysian. Ever since the Sidek's children and Lee Chong Wei played for internationals and made Malaysia renowned worldwide [1] Malaysians are interested to play in the evenings with their friends and family. Unfortunately, there are cheats and scandals occur in badminton by the players and the umpire .Nowadays, to judge a game fairly is challenging as human eyes has its limitation. Without proper equipment's or items, a game is biased especially the underhanded method between the players and the judge especially in line call. Such incident could spark anger to both supporters and players. Therefore, various methods had been set to create a fairly judged game without any human interference. This project is aimed to create an Automated Line Calling which is an automatic shuttlecock detector that can detect the shuttlecock when it falls on the line. This system is installed on the line of the court and each are embedded with force sensor which could differentiate between the feature given off by the player and the shuttlecock. When the shuttlecock hits on the line, the system will show an "IN" and "OUT" on the LCD display. Each corner of the court has a box shaped mechanism to detect the sensors which acts as receiver for the force detected by the sensor. The system has software which connects both hardware and software. Hence, when a product is created, the system is plugged in and automatically detect any pressure which is programmed in the software. Therefore, when it detect the force emitted by the shuttlecock, it will automatically display the "in "or "out". The result are verified by simulation and hardware result to prove the effectiveness of the propose product. Cyclops was created by an aeronautics engineer, Bill Carlton. This automatic line judge was the first to commercially introduce in the 1980s emphasise the usage of the beam system. There are transmitter and receiver boxes on the either sides of the net which transmit 5-6 infrared beams above 1cm from the ground. [2]The system was built so that one beam is

Content from this work may be used under the terms of the Creative Commons Attribution 3.0 licence. Any further distribution of this work must maintain attribution to the author(s) and the title of the work, journal citation and DOI.
Published under licence by IOP Publishing Ltd

IOP Conf. Series: Journal of Physics: Conf. Series **1020** (2018) 012011 doi:10.1088/1742-6596/1020/1/012011

on the "in" side while the other 4 beams are at the "out" side. If the ball hit the "in" beam line , the system automatically turned off the other 4 beams whereas when the ball hit either one of the 4 "out" beams ,the system produced a distinct beep. The systems had to be turned on before a serve and turned off after the serve which usually handled by the service line umpire. Figure 1 below shows the Cyclops system [2,3].

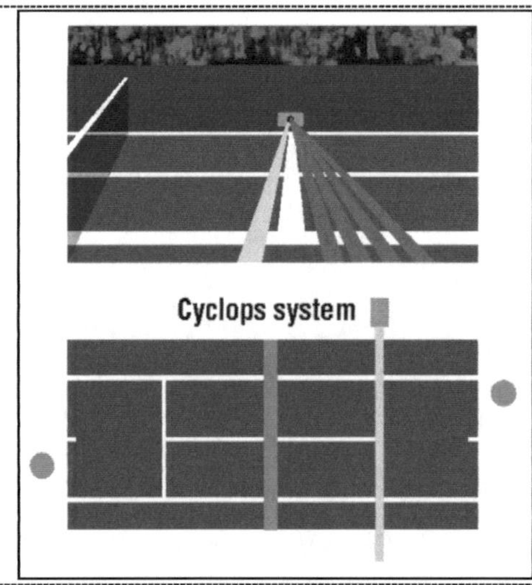

Figure 1: A system called the Cyclops uses infrared beams to determine whether a serve is in or otherwise

2. Methodology

This project is done on the flat area, whereas the velostat is used. The theory behind the function the velostat is, it is type of sensors that apply the principle of the increase of force cause the resistance to decrease over time. When the shuttle hit the velostat, the value of the velostat decreases. Technically, the parameter of a shuttlecock weighs around 4.75 to 5.50 grams (0.168 to 0.194 oz). The top 20 of the fastest smash in badminton recorded and the players' names, which was recorded with the speed guns in the tournament as tabulated in Table 1.

Table 1: Smashes speed of various players

No	Speed of smash, kph	Name of players
1	390	TAN WEE KIONG
2	393	GOH V SHEM
3	393	MADS KOLDING
4	395	IHSAN MAULANA MUSTOFA
5	395	GOH V SHEM
6	395	LEE CHONG WEI
7	396	FU HAIFFENG
8	397	CHAI BIAO
9	397	GOH V SHEM
10	397	LEE CHONG WEI
11	398	LEE CHONG WEI

12	399	ANTHONY GINTING
13	399	GOH V SHEM
14	399	LIN DAN
15	399	GOH V SHEM
16	400	HONG WEI
17	401	VLADIMIR IKANOV
18	402	GOH V SHEM
19	402	TANONGSAK SAENSOMBOONSUK
20	426	MADS KALDING

These smashes are recorded if they are more than 390 kph. The speed ranged from 390 to 426 kph as the 426 is the highest speed ever recorded in the tournament. From the table above it can be concluded that the mode for the speed of smash is 399 as it has been recorded for 4 times in the tournament. The median for the speed of the smash is

(390+393+393+395+395+395+396+397+397+397+398+399+399+399+399+400+401+402+402+42) /20 =398.5

Thus, the median force for the strongest smash detected based on the formula F=ma is 2.192 N.

3. Results and discussion

Since the output of the arduino operate by the voltage, it is important to understand the relation of force and voltage. The difference in voltage measured when moving from point A to point B is equal to the work which would have to be done, per unit charge, against the electric field to move the charge from A to B. When a voltage is generated, it is sometimes called an "electromotive force" or emf as shown equation 1 and 2.

$$emf = \frac{FL}{q} = \frac{qvBL}{q} \quad (1)$$

$$emf = vBL \quad (2)$$

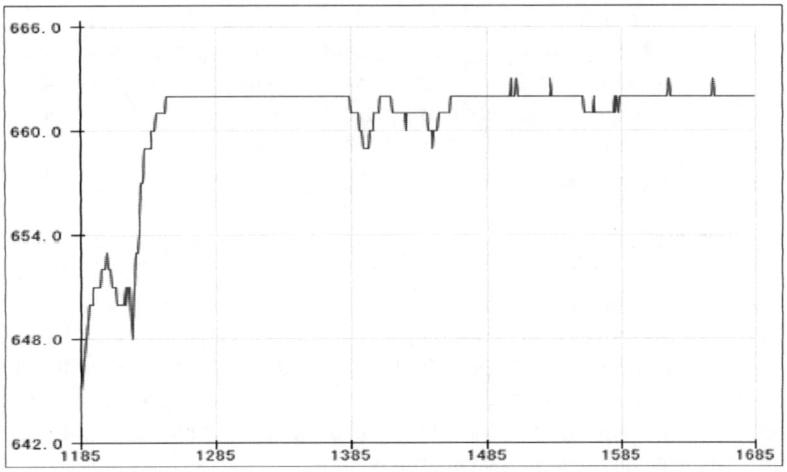

Figure 2: The startup of the sensor in terms of voltage over time

Figure 2 show the startup of the sensor in terms of voltage over time. The sensor is highly sensitive and detects small changes that may occur. Any vibrations or pressure applied will change the condition of the graph. To avoid the condition a fixed pressure offset is exerted on the sensor with a weight that has been placed at least half an hours before starting the test, this ensures that the material has reached the stability.

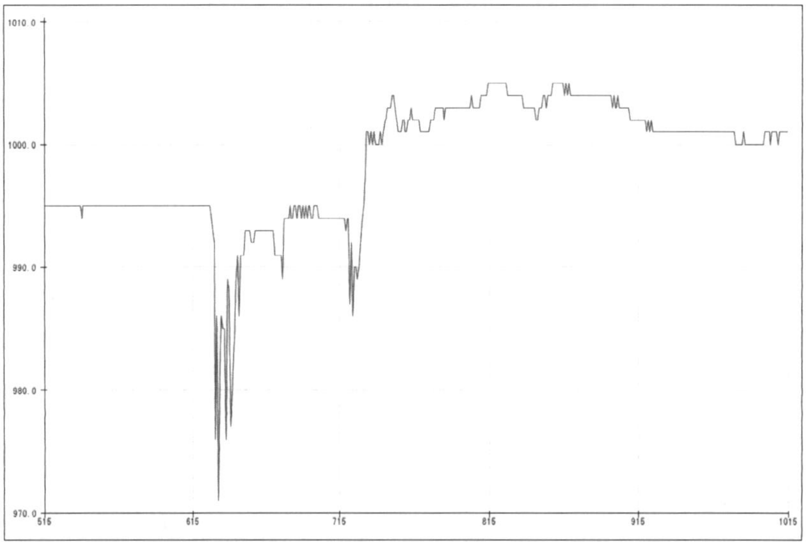

Figure 3: The impact on the velostat caused by the shuttle

Since the analog pin is code as the PULLUP pin whereas the voltage is continuosly apply across it, it will provide a different effect such as when the resistance is decrease when the force is applied. Figure 3 shows the impact on velostat casused by the shutte. The sharp drop of the above graph shows the impact on the velostat caused by the shuttle decrease the resistance across it. When the resistance drops, as stated in the ohm's law , the voltage decreases and the output is shown as above. Any sudden drops in the resistance of the graph above are caused by reducing the resistance . However,the result over time shows after certain repeatability, the voltage peak suddenly from its usual margin before the graph it maintain its linearlity.With these results it can be state that on average the sensor behavior is monotonic but the repeatability of results is low since for the same pressure up to a factor two of difference is observed between the lowest and the highest output value. It has to be remarked that these results have been obtained using the average of ten test cycles; therefore this processing technique is suitable for periodic signals only; moreover the output has a delay equal, in this case, to five times the cycle duration. Moreover error is calculated using the same test set used to calculate system coefficients; then it is an underestimation. From the test that had been done on this custom-made sensor it can show that the repeatability is one of the major issues. To try solving this applied a technique we called Period Averaging which helps to get more reliable results, but this has some limitations; for example, it can be applied only on signals which are known to have some form of periodicity. The low accuracy of single point sensors, the cost and complexity of sensor matrices, and issues related to measure pressure are certainly some key factors for the relatively low diffusion of this technology for commercial devices. The result or the score of the game may be able to achieve, but the time taken for the velostat to reach its stability is time consuming as it takes certain time to reach its stability. A certain calculation technique such as called Period Averaging is proposed to limit uncertainty issue but it is not resolutive. Considering that the time needed to review the ball falling on the line may take some time and may be considered as fault.

IOP Conf. Series: Journal of Physics: Conf. Series **1020** (2018) 012011 doi:10.1088/1742-6596/1020/1/012011

4. Conclusion

As a conclusion, in this state that measure absolute pressure with this sensing technology could be very challenging, but for other applications where spatial information is the key, they can be used exploiting their low cost and simplicity.

References

[1] "Khairy says badminton is now Malaysia's number one sport", Themalaymailonline.com, 2017. [Online]. Available: http://www.themalaymailonline.com/sports/article/khairy-says-badminton-is-now-malaysias-number-one-sport. [Accessed: 18- May- 2017].

[2] G. F. Grant, "Tennis Court Line Monitoring Apparatus." U.S. Patent 3982759 A, issued Sept. 28, 1976.

[3] Thomas Cup - Badminton", Badminton-information.com, 2017. [Online]. Available: http://www.badminton-information.com/thomas_cup.html. [Accessed: 18- May- 2017].

[4] "History of Badminton", Badminton-information.com, 2017. [Online]. Available: http://www.badminton-information.com/history-of-badminton.html. [Accessed: 18- May- 2017].

IOP Conf. Series: Journal of Physics: Conf. Series **1020** (2018) 012012 doi:10.1088/1742-6596/1020/1/012012

Material selection in a sustainable manufacturing practice of a badminton racket frame using Elimination and Choice Expressing Reality (ELECTRE) Method

Muhammad Firdhaus Che Hassan, Mohd Uzair Mohd Rosli and Muhammad Afiq Mohd Redzuan

Department of Mechanical Engineering Technology, Faculty of Engineering Technology, Universiti Malaysia Perlis 02100 Padang Besar, Perlis, Malaysia

firdhaus@unimap.edu.my / muhammadfirdhauschehassan@gmail.com

Abstract. Badminton is one of the leading sports in the world. It has its own set of rules on the equipments used and general game play. One of the main equipment used is the badminton racket. Each sections of a badminton racket have its own design requirements and one of it is the racket's material selection. Therefore, material selection is very important to improve the usage of a badminton racket. This study describes the use of the Elimination and Choice Expressing Reality (ELECTRE) method in the material selection of a badminton racket frame with reference to the sustainable manufacturing practice of the frame. By categorizing the materials of the badminton racket frame according to mechanical, physical, chemical and environmental properties, and further detailed sub criteria were set according to the usage of these frames, the ELECTRE I method was used to determine the dominant material. Out of the six materials usually used in the manufacturing of a badminton racket frame, carbon fibre was the dominant material selected from three out of the four properties which are the mechanical, chemical and most importantly the environmental properties, as to comply with the sustainable manufacturing practice of these frames.

1. Introduction

1.1. Background

In the sports world now there are different tools needed to perform the activities and develop and innovate it up to date. Increasing sophisticated technology in sports has contributed a lot to increase the usage of a sports equipment. The design requirements of a sports equipment is an important aspect and commonly associated with conceptual design, material selection and process selection in terms of principles of design and technology. This is especially to produce new material that improves the quality of the sports equipment specifically and the game itself generally [1]. The basic reason to produce a badminton racket is to improve the quality and use of the badminton racket. Factors that needed to be improved are the quality of materials and optimizing the design of a badminton racket [2].

Content from this work may be used under the terms of the Creative Commons Attribution 3.0 licence. Any further distribution of this work must maintain attribution to the author(s) and the title of the work, journal citation and DOI.
Published under licence by IOP Publishing Ltd

IOP Conf. Series: Journal of Physics: Conf. Series **1020** (2018) 012012 doi:10.1088/1742-6596/1020/1/012012

To design is to apply the latest methods and ideas. Each design stage requires a decision with regard to materials in the manufacture of products and methods needed to produce it. The selection of material is important in the production of a product. It is a key element in developing a product because it is an impact or effect on a product developed. Material choice is a step inside the process of designing any physical item. In the context of product design, the principle intention of material selection is to minimize value even as meeting product performance desires [3].

In the manufacturing industry, any equipment which is derived require proper attention to avoid any losses occurred as a loss in the use of raw materials to produce a product. Therefore, steps should be taken to perform the sustainability in a manufacturing point of emphasis that whereby in every material used will not be wasted. According to the definition from the United States Environmental Protection Agency, sustainable manufacturing is the creation of a manufactured product with processes that have minimal negative impact on the environment, conserve energy and natural resources, are safe for employees and communities, and are economically sound [4].

The focus of this study was to determine the criteria and sub criteria of the material properties of a badminton racket frame, use ELECTRE I method to apply material selection thus the dominant material will be chosen.

1.2. Badminton Racket Frame

A badminton racket consists of a head, a shaft that is joined to the head, and it is between the head and a handle. Altogether it is called a frame. The head also holds strings to make it feasible to play with [5]. All of these parts are shown in figure 1 below.

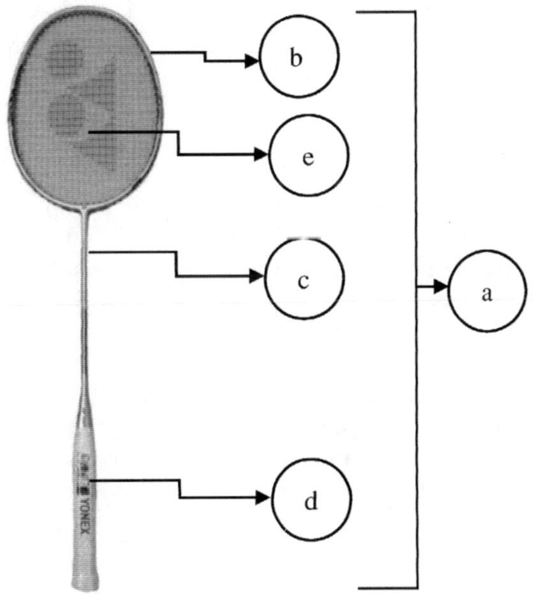

Figure 1: Badminton racket - a) frame, b) head, c) shaft, d) handle and e) strings.

1.3. Materials in a Badminton Racket Frame

The following are the materials chosen to be included in this material selection process using ELECTRE I method.

1.3.1. Carbon Fibre Carbon fibre is made from organic fibre classics carbonization and graphitization processing. Carbon fibre consists of carbon that is 90% higher than that of inorganic polymer, consisting of 99% carbon graphite fibre [6].

IOP Conf. Series: Journal of Physics: Conf. Series **1020** (2018) 012012 doi:10.1088/1742-6596/1020/1/012012

1.3.2. Boron Fibre Boron fibre is a wire of tungsten with elemental boron deposited during a reaction of boron trichloride and hydrogen gas at 800°C to 2,000°C. Boron fibre is so stiff and strong that it has been used to patch across the fuselage of jet fighter aircraft to hold the wings on [7].

1.3.3. Aluminium Primary aluminium is aluminium tapped from electrolytic cells or pots during the electrolytic reduction of metallurgical alumina (aluminium oxide – Al2O3). It thus excludes alloying additives and recycled aluminium. [8].

1.3.4. Carbon Steel Carbon steel is the most extensively used materials in the industry. This material is very strong and holds shape memory well, making it ideal for springs and wire. Carbon steel is a composite comprising of iron and carbon [9].

1.3.5. Nickel Titanium Nickel titanium (NiTi) is a combination of two different metals which are nickel and titanium metals, where these elements have the same percentage of atoms. Nickel titanium was developed 40 years ago in the Naval Ordnance Laboratory (NOL) in Silver Springs, Maryland thus creating the acronym NiTiNOL, which is used worldwide for this special type of alloy [10].

1.3.6. Titanium Alloy Titanium alloys are widely used in the engineering field, namely in the aerospace, automotive and biomedical parts, because of their high specific strength and exceptional corrosion resistance [11].

1.4. Sustainable Manufacturing

Sustainable manufacturing has been adulated as of late for its huge advantages coordinated at triple main concern factors (social, economy, and financial), but the majority of manufacturing strategies remain limited to either one or two factors. No dependable rules exist to ensure the fruitful usage of feasible assembling; accessible writing neglects to investigate such an unpredictable research theme [12].

1.5. Elimination and Choice Expressing Reality (ELECTRE)

Elimination and Choice Expressing Reality or ELimination Et Choix Traduisant la REalité (ELECTRE) methods have a variety of methods such as of ELECTRE have been produced including ELECTRE I, II, III, IV and TRI. All of these techniques are the same, but different in terms of operation. Each technique based on the same basic concept but its operation is different. The important thing of ELECTRE approach is the utilization of the idea of outranking relationship [13]

ELECTRE I is to determine higher-degree relations between exclusive bid schemes, thru accordance and disaccord examine, and seek for the non-inferior scheme subset unceasingly, in the end, decide the first-class bid scheme. This approach considered the actual situation and the fee preference of creation units and overcame the supplementary state of affairs among evaluative indicators, and it's going to offer a brand-new approach foundation for the selection-makers [14].

2. Methodology

2.1. Material Properties (Criteria) and Sub Criteria Identification

The material properties or criteria and its sub criteria were chosen based on the function or usage of the badminton racket frame. Its function includes game playing, storing at home or travelling with in transports. These provide clues as to which criteria and sub criteria to choose from and applied in the calculations of ELECTRE I.

2.1.1. Mechanical Properties This was chosen with the sub criteria of Ultimate Tensile Strength (Mpa), Elastic Modulus (Gpa), Poisson's Ratio and Rockwell Hardness in mind, all relating to the game play effect on the badminton racket frame.

IOP Conf. Series: Journal of Physics: Conf. Series **1020** (2018) 012012 doi:10.1088/1742-6596/1020/1/012012

2.1.2. Physical Properties The following sub criteria was identified with game play, storing and travel use in mind - Density (g/cm^3), Thermal Conductivity (W/m-K), Melting Point (°C) and Hardness (Mpa).

2.1.3. Chemical Properties The sub criteria of Corrosion Resistant, Heat of Combustion (kJ/kg) and Wear Resistance are more into the storing and travel usage in mind.

2.1.4. Environmental Properties These sub criteria are critical as it is related to the sustainable manufacturing of this product, and it is the following: Energy Content (MJ/kg), CO2 Emission (g) and Flammability.

2.2. ELECTRE I

The ELECTRE I technique is basically connected to the treatment of discrete options esteemed quantitatively, to a fractional appointment of choices. The point is to isolate from the entire of the choices the ones that are favoured on most of the assessment criteria and which don't bring about inadmissible level of discontent on other criteria.

The establishment of a relation of preference R makes the comparison between alternatives not necessarily transitive. For example: *a1*, *a2* and *a3* are the different alternatives. If *a1 R a2*, it means that *a1* is preferable to *a2*. However, if *a2 R a3* does not mean that *a1 R a3*. So, in the determination of the relation of preference *R*, the concept of agreement and disagreement are very important to making a good decision.

The agreement between two alternatives, *i* and *j*, represent the will of the decision maker in choosing alternative *i* over alternative *j*. The agreement calculation $i = \{1, 2, 3, \dots, n\}$ represents the set of *n* criteria evaluation, and $\{a_k : k = 1, 2, 3, \dots, n\}$ represent the set of weights associated to *n* criteria. These weights are determined by the judgment of value of the decision maker.

The following notation is used:

i is preferable to *j*: $i > j$

i is equally preferable to *j*: $i = j$

2.2.1. Index of Agreement & Index of Disagreement The index of agreement may be understood as a pondered percentage of the calibre to which the alternative *i* is preferred to alternative *j*. By definition, $0 \leq C(i, j) \leq 1$. It is convenient to present the indexes of agreement in an agreement matrix, C, where $C\ (i, j)$ is the element of row i and column j. The index of agreement is meant as:

$$C\ (i, j) = \frac{(W^+ + 0.5W^=)}{(W^+ + W^= + W^-)}$$

This is whereby $C\ (i,\ j)$ are component of matrix C that can be structured with component of row i and column j. W^+ is the positive weightage which follow the rule of Index of Agreement, $W^=$ is the equal weightage and W^- is the negative weightage which do not follow the rule of Index of Agreement.

The meaning of disagreement is complementary to the one of agreement and represents the discomfort experienced in the choosing of alternative *i* above alternative *j*. In the order to calculate the index of disagreement it is meaning a numerical scale common to all calibre. This scale is used to compare of make a different the discomfort, to combine with all calibre, cause by the choosing of alternative *i* instead of alternative *j*. By these considerations, the index of disagreement is meant as:

$$D\ (i, j) = \max k \in I \frac{[z(j, k) - z(i, k)]}{R^*}$$

IOP Conf. Series: Journal of Physics: Conf. Series **1020** (2018) 012012 doi:10.1088/1742-6596/1020/1/012012

This is whereby $D\ (i, j)$ is the component of matrix D that can be structured with component of row i and column j, $z\ (j, k)$ is the evaluation of option j, based on the numerical scale of model k, and R^* is the higher prevalent value of numerical scales.

Where $z(j,k)$ is the evaluation of alternative j, according to the numerical scale of the criterion k, and R^* is the highest superior value of the numerical scales. A disagreement matrix D can be formation, where $D(i,j)$ is the element of row i and column for j.

The connection of preference R is meaning when organize which by the Decision Maker as far as values (p,q), between zero and one, so as alternative i is contrived to a j if and only if the $C(i,j) \geq p$ and $D(i,j) \leq q$. (usually $p \geq 0.75$ and $q \leq 0.2$)

2.2.2. ELECTRE Graph and Kernel's Determination Inside this graph, every node or hub is shown as a circled number. Every node compares to a choice. For example, there are 4 choices or option in the set. The arrows exuding from the nodes are called coordinated ways and compare to an outranking connection. Here option number 4 outranks 1, 2 and 3, and 1 outranks 2. The bit of the diagram in this way comprises of four options and is the subset of non-overwhelmed options which the ELECTRE I method characterizes.

In the ELECTRE I method, the set of alternatives in consideration is reduced by determining of a subset denominated as Kernel, *K*, and defined by:

a) No system in *K* dominates another system in also in *K*
b) Each system outside Kernel is dominated by at least in *K*

Kernel contains the good fine system based in *R*. The systems outside Kernel are eliminated from future considerations. The connections developed by ELECTRE I for this situation can be promptly communicated as a diagram such as in figure 2 below, knowing that by applying Kernel's Determination, option number 4 is dominant and chosen.

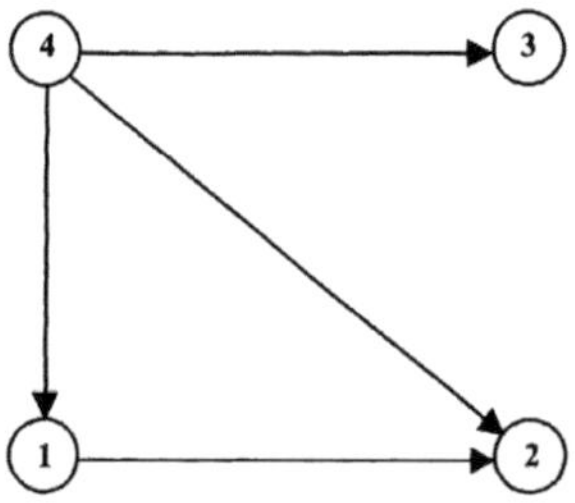

Figure 2: The ELECTRE graph

3. Results

3.1. Material Properties (Criteria) and Sub Criteria Value Identification
The values of all sub criteria was searched and identified from sources such as handbooks and material yearly reports to get the value usually used for these sub criteria towards the materials chosen.

3.1.1. Mechanical Properties Values Table 1 below shows the Mechanical Properties' sub criteria values.

Table 1: Mechanical properties values.

Criteria	Sub Criteria	Materials					
		Carbon Fibre	Boron Fibre	Aluminium Alloy 7000	High Carbon Steel	Nickel Titanium 60	Titanium Alloy 6Al-4V
Mechanical	Ultimate	4127	3500	600	1750	1172	950

IOP Conf. Series: Journal of Physics: Conf. Series **1020** (2018) 012012 doi:10.1088/1742-6596/1020/1/012012

Properties	Tensile Strength (Mpa)						
	Elastic Modulus (Gpa)	300	40	79	210	110	113.8
	Poisson's Ratio	0.28	0.27	0.34	0.33	0.33	0.34
	Rockwell Hardness (HR)	57	40	49.8	70	62	36

3.1.2. Physical Properties Values Table 2 below shows the Physical Properties' sub criteria values.

Table 2: Physical properties values.

Criteria	Sub Criteria	Materials					
		Carbon Fibre	Boron Fibre	Aluminium Alloy 7000	High Carbon Steel	Nickel Titanium 60	Titanium Alloy 6Al-4V
Physical Properties	Density (g/cm^3)	2.22	2.57	2.82	7.85	6.45	4.43
	Thermal Conductivity (W/m-K)	24	45	180	52	18	7.1
	Melting Point (°C)	3550	2040	654	1400	1310	1605
	Hardness (Mpa)	680	550	60	164	667	3370

3.1.3. Chemical Properties Values Table 3 below shows the Physical Properties' sub criteria values.

Table 3: Chemical properties values.

Criteria	Sub Criteria	Materials					
		Carbon Fibre	Boron Fibre	Aluminium Alloy 7000	High Carbon Steel	Nickel Titanium 60	Titanium Alloy 6Al-4V
Chemical Properties	Corrosion Resistant (1=Poor/ 5=Excellent)	5	4	4	4	4	5
	Heat of Combustion (kJ/kg)	3463	507	3098	715	378	425
	Wear Resistance (1=Poor 5=Excellent)	4	3	3	3	4	3

3.1.4. Environmental Properties Values Table 4 below shows the Environmental Properties' sub criteria values.

Table 4: Environmental properties values.

Criteria	Sub Criteria	Materials					
		Carbon Fibre	Boron Fibre	Aluminium Alloy 7000	High Carbon Steel	Nickel Titanium 60	Titanium Alloy 6Al-4V
Chemical Properties	Energy Content (MJ/kg)	103	200	42.5	9.8	0.77	750
	CO2 Emission (g)	44.0	$15x10^5$	600	4.0	$2.2x10^7$	5.2
	Flammability (1=Poor 5=Excellent)	3	4	5	4	5	5

IOP Conf. Series: Journal of Physics: Conf. Series **1020** (2018) 012012 doi:10.1088/1742-6596/1020/1/012012

3.2. ELECTRE I Calculations

For the next step, the calculations of ELECTRE I will be applied and only the Mechanical Properties values will be shown.

3.2.1. Index of Agreement and Index of Disagreement for Mechanical Properties

The first step is to calculate the Index of Agreement. For this step, decide on a weightage for each sub criteria based on the Decision Maker's Preferences: Ultimate Tensile Strength is 5, Elastic Modulus is 3, and the values for Poisson's Ratio and Rockwell Hardness are 1. All values used must be enough and no more than 10. Table 5 below shows the weights attributed to all sub criteria of mechanical criteria.

Table 5: Weightage of the sub criteria in the mechanical properties criteria.

Sub Criteria	Weightage
Ultimate Tensile Strength	5
Elastic Modulus	3
Poisson's Ratio	1
Rockwell Hardness	1
Total Weightage	10

Each sub-criterion should have its own rank and code values. For these mechanical criteria, each sub criterion has six fractional value levels. For the Ultimate Tensile Strength sub criteria, the values of level used are divided are into: than 4100 for code A, range from 3501 to 4100 for code B, 1800 to 3500 for code C, 1200 to 1799 for code D, 701 to 1199 for code E and 0 to 700 for code F. For the Elastic Modulus sub criteria, the values of level used are divided into: more than 250 for code G, range from 200 to 250 for code H, 150 to 199 for code I, 100 to 149 for code J, 50 to 99 for code K and 0 to 49 for code L. As for Poisson's Ratio sub criteria, it is also divided into six (6) values of sub criteria which are: more than 0.35 for code M, range 0.29 to 0.35 for code N, 0.23 to 0.28 for code O, 0.17 to 0.22 for code P, 0.11 to 0.16 for code Q and 0 to 0.10 for code R. And last of the sub criteria of mechanical properties is Rockwell Hardness, it is divided into: more than 70 for code S, range from 58 to 69 for code T, 46 to 57 for code U, 35 to 45 for code V, 23 to 34 for code W and 0 to 22 for code X. All of these data are shown in table 6 below.

Table 6: The levels and codes of sub criteria in mechanical properties criteria

Sub Criteria	Level	Code
Ultimate Tensile Strength (Mpa)	>4100	A
	3501-4100	B
	1800-3500	C
	1200-1799	D
	701-1199	E
	0-700	F
Elastic Modulus (Gpa)	>250	G
	200-250	H
	150-199	I
	100-149	J
	50-99	K
	0-49	L
Poisson's Ratio	>0.35	M
	0.29-0.35	N
	0.23-0.28	O
	0.17-0.22	P

	0.11-0.16	Q
	0-0.10	R
Hardness Rockwell	>70	S
	58-69	T
	46-57	U
	35-45	V
	23-34	W
	0-22	X

Next step is spread out the matrix of the material selection evaluation of the badminton racket frame for the mechanical criteria. Each of the materials is classified to numbers as coded from 1 to 5 and as follows; (1) Carbon Fibre, (2) Boron Fibre, (3) Aluminium Alloy 7000 series, (4) High Carbon Steel, (5) Nickel Titanium 60, and (6) Titanium Alloy 6Al-4V. The table 7 below shows the matrix of material selection of the badminton racket frame for the Mechanical Properties criteria.

Table 7: Matrix of the material selection evaluation for mechanical properties criteria.

Sub-criteria	Material					
	1	2	3	4	5	6
Ultimate Tensile Strength (Mpa)	A	B	E	C	D	E
Elastic Modulus (Gpa)	G	L	K	H	I	J
Poisson's Ratio	N	O	N	N	N	N
Rockwell Hardness	T	V	T	S	T	V

Now, the calculation can be with done for index of agreement. Below shows an example of calculation for Index of Agreement for Mechanical Properties between material 1 and 2.

$$C(1,2) = \frac{1}{10}(5 + 3 + 0 + 1) = 0.9$$

Do this for all possible materials contradictions for which every answer is then combined in a matrix form, for it to become the Index of Agreement for Mechanical Properties, as below.

$$C(Mechanical\ Properties) = \begin{bmatrix} - & 0.9 & 0.9 & 1 & 0.85 & 0.95 \\ 0.1 & - & 0.1 & 0.5 & 0.05 & 0.2 \\ 0.1 & 0.9 & - & 0.5 & 0.05 & 0.95 \\ 0 & 0.5 & 0.5 & - & 0.5 & 0.55 \\ 0.15 & 0.95 & 0.95 & 0.45 & - & 0.05 \\ 0.05 & 0.4 & 0.05 & 0.45 & 0.05 & - \end{bmatrix}$$

Next, make the calculations Index of Disagreement for Mechanical Criteria. For Index of Disagreement, use the Decision Maker's Preferences of Numerical Scale Value. Decide the Numerical Scale Value for each sub criteria of Mechanical Criteria. The maximum Value of Numerical Scale is 100. Decide the Numerical Values on each sub criteria. The sub criteria's numerical values: Ultimate Tensile Strength is 100, Elastic Modulus is 70, Poisson's Ratio is 50 and the Rockwell Hardness value is 40, as shown in table 8 on the next page

IOP Conf. Series: Journal of Physics: Conf. Series **1020** (2018) 012012 doi:10.1088/1742-6596/1020/1/012012

Table 8: Maximum values of numerical scales in the mechanical properties criteria.

Sub Criteria	Numerical Scale Value
Ultimate Tensile Strength (Mpa)	100
Elastic Modulus (Gpa)	70
Poisson's Ratio	50
Hardness Rockwell	40

Then, arrange and determine the codes and Numerical Scale Value for Index of Disagreement for each Numerical Scale Value of criteria as in table 9.

Table 9: Values determined for the Index of Disagreement in the mechanical properties.

Sub Criteria	Numerical Scale Value	Code
Ultimate Tensile Strength (Mpa)	100	A
	83	B
	66	C
	49	D
	32	E
	15	F
Elastic Modulus (Gpa)	70	G
	59	H
	48	I
	37	J
	26	K
	15	L
Poisson's Ratio	50	M
	42	N
	34	O
	26	P
	18	Q
	10	R
Hardness Rockwell	40	S
	33.5	T
	27	U
	20.5	V
	14	W
	7.5	X

Then, calculate all sub criteria for the Index of Disagreement for Mechanical Properties. For this calculation, use the same Matrix of the Material Selection Evaluation on table 7 for reference. Calculate each sub criteria one by one and then compile all the data in four (4) different matrices which are; Matrix D for Ultimate Tensile Strength sub criteria, Matrix D for Elastic Modulus sub criteria, Matrix D for Poisson's Ratio sub criteria and Matrix D for Rockwell Hardness. The calculation of sub criteria for Ultimate Tensile Strength between material 2 and 1 is shown below. If a negative (-) value was calculated, automatically it will become a zero value.

$$D(2,1) = \frac{(100-32)}{100} = 0.68$$

IOP Conf. Series: Journal of Physics: Conf. Series **1020** (2018) 012012 doi:10.1088/1742-6596/1020/1/012012

In this step, determine the calculations in the Index of Disagreement for all four (4) sub criteria.

$$D(Ultimate\ Tensile\ Strength) = \begin{bmatrix} - & 0 & 0 & 0 & 0 & 0 \\ 0.68 & - & 0.17 & 0.51 & 0.34 & 0 \\ 0.51 & 0 & - & 0.34 & 0.17 & 0 \\ 0.17 & 0 & 0 & - & 0 & 0 \\ 0.34 & 0 & 0 & 0.17 & - & 0 \\ 0.68 & 0 & 0.17 & 0.51 & 0.34 & - \end{bmatrix}$$

$$D(Elastic\ Modulus) = \begin{bmatrix} - & 0 & 0 & 0 & 0 & 0 \\ 0.16 & - & 0 & 0 & 0 & 0 \\ 0.31 & 0.16 & - & 0 & 0 & 0 \\ 0.19 & 0.03 & 0 & - & 0 & 0 \\ 0.6 & 0.47 & 0.31 & 0.16 & - & 0 \\ 0.79 & 0.63 & 0.47 & 0.31 & 0.16 & - \end{bmatrix}$$

$$D(Poisson's\ Ratio) = \begin{bmatrix} - & 0 & 0 & 0 & 0 & 0 \\ 0 & - & 0 & 0 & 0 & 0 \\ 0 & 0 & - & 0 & 0 & 0 \\ 0.16 & 0.16 & 0.16 & - & 0.16 & 0.16 \\ 0 & 0 & 0 & 0 & - & 0 \\ 0 & 0 & 0 & 0 & 0 & - \end{bmatrix}$$

$$D(Hardness\ Rockwell) = \begin{bmatrix} - & 0 & 0 & 0 & 0.16 & 0 \\ 0 & - & 0 & 0 & 0.16 & 0 \\ 0 & 0 & - & 0 & 0.16 & 0 \\ 0.33 & 0.33 & 0.33 & - & 0.49 & 0 \\ 0 & 0 & 0 & 0 & - & 0 \\ 0.33 & 0.33 & 0.33 & 0 & 0.49 & - \end{bmatrix}$$

Make comparison and combine all sub criteria by maintaining the highest value only. Below shows the matrix form for all sub criteria of Mechanical Properties.

$$D(Mechanical\ Criteria) = \begin{bmatrix} - & 0 & 0 & 0 & 0 & 0 \\ 0.68 & - & 0.17 & 0.51 & 0.34 & 0 \\ 0.51 & 0.16 & - & 0.34 & 0.17 & 0 \\ 0.33 & 0.33 & 0.33 & - & 0.49 & 0.16 \\ 0.34 & 0.47 & 0.31 & 0.17 & - & 0 \\ 0.79 & 0.63 & 0.47 & 0.33 & 0.49 & - \end{bmatrix}$$

Then, compare between the Index of Agreement (matrix C) and Index of Disagreement (matrix D). To make comparison, find the correlation using the rules. Below show the rules for comparison.

$C(i,j) \geq p\ where; p \geq 0.75$ (for Index of Agreement)
$D(i,j) \leq q\ where; q \leq 0.20$ (for Index of Disagreement)

From the matrices, value for the index of agreement must be 0.75 and above. Meanwhile, the value of disagreement must be 0.20 and below.

$$C(Mechanical\ Criteria) = \begin{bmatrix} - & 0.9 & 0.9 & 1 & 0.85 & 0.95 \\ 0.1 & - & 0.1 & 0.5 & 0.05 & 0.2 \\ 0.1 & 0.9 & - & 0.5 & 0.05 & 0.95 \\ 0 & 0.5 & 0.5 & - & 0.5 & 0.55 \\ 0.15 & 0.95 & 0.95 & 0.45 & - & 0.05 \\ 0.05 & 0.4 & 0.05 & 0.45 & 0.05 & - \end{bmatrix}$$

IOP Conf. Series: Journal of Physics: Conf. Series **1020** (2018) 012012 doi:10.1088/1742-6596/1020/1/012012

$$D(Mechanical\ Criteria) = \begin{bmatrix} - & 0 & 0 & 0 & 0 & 0 \\ 0.68 & - & 0.17 & 0.51 & 0.34 & 0 \\ 0.51 & 0.16 & - & 0.34 & 0.17 & 0 \\ 0.33 & 0.33 & 0.33 & - & 0.49 & 0.16 \\ 0.34 & 0.47 & 0.31 & 0.17 & - & 0 \\ 0.79 & 0.63 & 0.47 & 0.33 & 0.49 & - \end{bmatrix}$$

From this matrix, the values highlighted have been determined to comply with the rules set previously. Then, determine the same row and column to get the chosen materials contradictions from Matrix C and Matrix D. From these matrices, the materials that coincides are (1, 2), (1, 3), (1, 4), (1, 5), (3, 2) and (3, 6).

3.2.2. ELECTRE Graph for Mechanical Properties From the matrices, the indices identified for both sets are (1, 2), (1, 3), (1, 4), (1, 5), (3, 2) and (3, 6). ELECTRE graph can be constructed with using the chosen matrices. Figure 3 below shows the ELECTRE graph for Mechanical Properties criteria.

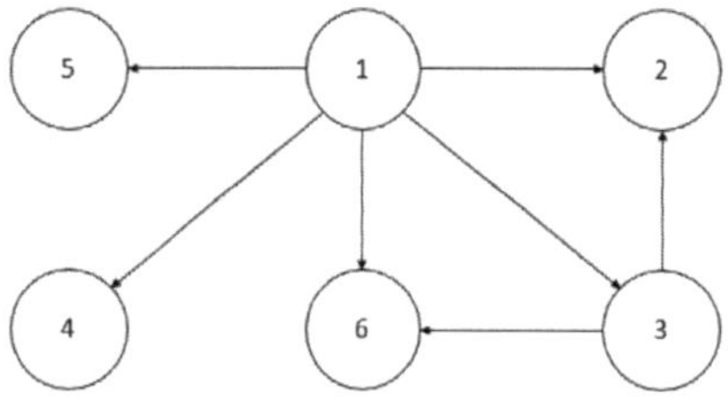

Figure 3: ELECTRE graph for mechanical properties criteria.

3.2.3. Kernel's Determination for Mechanical Properties Criteria The Kernel's Determination can be done by selecting the dominant node. From figure 3 above, only node number 1 satisfies this condition. From this graph, the material to be considered is material 1. So, the best material to be chosen for a badminton racket frame based on Mechanical Properties criteria is material 1 which is Carbon Fibre.

3.2.4. Repetition of Calculations The procedure from subsubsections 3.2.1 to 3.2.3 was repeated to get the dominant material to be chosen for Physical Properties, Chemical Properties and Environmental Properties criteria. The final ELECTRE graphs for all these criteria are shown below.

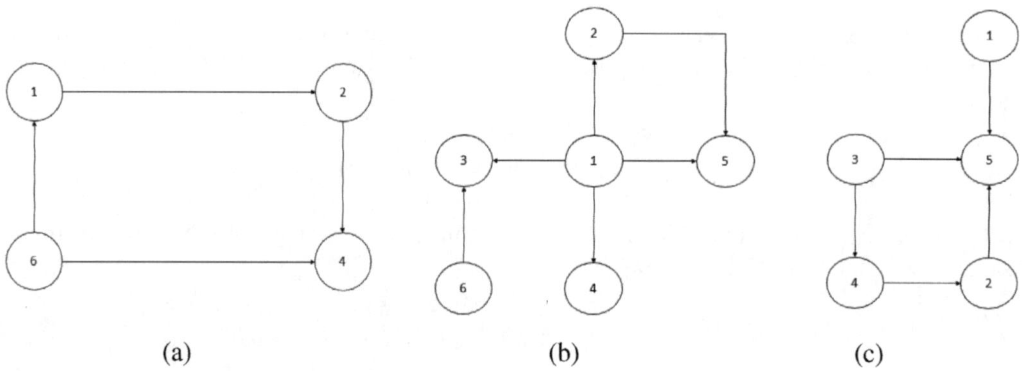

Figure 4: ELECTRE graph of a) physical properties, b) chemical properties and c) environmental properties criteria.

IOP Conf. Series: Journal of Physics: Conf. Series **1020** (2018) 012012 doi:10.1088/1742-6596/1020/1/012012

4. Discussion

4.1. Materials, Criteria and Sub Criteria Identifications

For this research, the chosen materials were the ones usually used as for now. Due to data restrictions from the industry and other sources, the criteria and sub criteria chosen was the ones that can get hold of. Some of them are the ones that are useful. Some has been changed to become a numerical data as it is more considerate to categorize them.

4.2. Applications

The method used, which was ELECTRE, can be applied in the industry of badminton racket frame production. Materials with the best characteristics can be identified and chosen in the conceptual design stage by using this method. From here, the customer or user can also identify which ones suit their playing style and prepare in advanced on storage and travelling usage of the badminton rackets.

5. Conclusion and Recommendation

5.1. Conclusion

The ELECTRE I method have chosen the dominant material of 1, which is Carbon Fibre as the material to be used in a sustainable manufacturing practice of a badminton racket frame. This is because it became dominant in Mechanical, Chemical and Environmental Properties criteria, with only the Physical Properties criteria not able to be dominant.

5.2 Recommendation

Industrial data sharing is a must to get the best materials, criteria and subcriteria to be used as part of the ELECTRE I calculations, whereby the locations of the production factories are all outside of Malaysia. This is to get the most accurate dominant material in a material selection process of a sustainable manufacturing practice of a badminton racket frame, and maybe more of the other parts too.

6. Acknowledgments

Authors wished to acknowledge assistance and support from Universiti Malaysia Perlis by giving out a Short Term Grant (9001-00518) to complete this research.

7. References

[1] Muller M, Senner V and Lindemann U 2007 Specific characteristics of sports equipment *Proc. Int. Conf. on Eng. Des. (ICED '07) (Paris)*

[2] Nasruddin F A, Syahrom A, Harun M N, Abdul Kadir M R and Omar A H 2014 Finite-element study on effect of string tension toward coefficient of restitution of a badminton racket string-bed *Adv. Mater. Res.* **845** 417-420

[3] Ashby M F 2005 *Materials Selection in Mechanical Design* (Oxford: Elsevier Butterworth-Heinemann)

[4] Quadrennial Technology Review: An Assessment of Energy Technologies and Research Opportunities 2015 *Chapter 6: Innovating Clean Energy Technologies in Advanced Manufacturing*

[5] Nasruddin F A, Syahrom A, Harun M N, Abdul Kadir M R, Omar A H and Öchsner A 2016 *Finite Element Analysis on Badminton Racket Design Parameters* ed H Altenbach *et al* (SpringerBriefs in Computational Mechanics) chapter 1 pp 1–14

[6] Zhang D, Liu H and Sun L 2015 Dynamic Analysis of Carbon Fiber-Reinforced Wood Composites Based on Finite Element Model *BioResources* **11(1)**

[7] Kostick D S 2007 *U.S. Geological Survey Minerals Yearbook-2006: Boron* (U.S. Geological Survey Publications)
[8] International Aluminium Institute 2015 *Primary Aluminium Production Reporting Guidelines* (London)
[9] Popović O, Prokić-Cvetković R, Sedmak A, Grabulov V, Burzić Z, Rakin M 2010 Characterisation of high-carbon steel surface welded layer *Strojniški vestnik – J. of Mech. Eng.* **56(5)** 295-300
[10] Ferreira M D A, Luersen M A and Borges P C 2012 Nickel-titanium alloys: A systematic review *Dent. Press J. of Orthodontics* **17(3)**
[11] Veiga C, Davim J P and Loureiro A J R 2013 Review on machinability of titanium alloys: The process perspective *Rev. Adv. Mater. Sci.* **34** 148-164
[12] Madan Shankar K, Kannan D and Udhaya Kumar P 2017 Analyzing sustainable manufacturing practices - A case study in Indian context *J. of Cleaner Prod.* **164** 1332-43
[13] Rogers M, Bruen M and Maystre L Y 2011 *ELECTRE and decision support: Methods and applications in engineering and infrastructure investment* (New York: London: Springer)
[14] Wu Y, Huang Y and Chen W 2011 Construction project bid evaluation optimization model based on the method of ELECTRE - I *Proc. 2011 IEEE 18th Int. Conf. on Ind. Eng. and Eng. Man. (Changchun)* (Beijing: Institute of Electrical and Electronics Engineers, Inc.) p 1660-63